Statistics

Third Edition

by Bob Donnelly Jr., PhD and Fatma Abdel-Raouf, PhD

A member of Penguin Random House LLC

Publisher: Mike Sanders
Associate Publisher: Billy Fields
Acquisitions Editor: Jan Lynn
Development Editor: Alexandra Elliott
Cover Designer: Lindsay Dobbs
Book Designer: William Thomas
Compositor: Ayanna Lacey
Proofreader: Laura Caddell
Indexer: Tonya Heard

To my wife, Debbie, who supported and encouraged me every step of the way. I could not have done this without you, babe.—Bob Donnelly Jr.

To my shining stars, who shine on me and brighten my day, and whom I love endlessly and unconditionally. You are my inspiration and motivation for writing this book.—Fatma Abdel-Raouf, PhD

First American Edition, 2016
Published in the United States by DK Publishing
6081 E. 82nd Street, Indianapolis, Indiana 46250

16 17 18 19 10 9 8 7 6 5 4 3 2 1
003–292192–July/2016

Published in the United States by Dorling Kindersley Limited.

ISBN: 978-1-46545-166-8
Library of Congress Catalog Card Number: 2015956731

Printed and bound in the United States of America

idiotsguides.com

Contents

Appendixes

Foreword

Statistics, statistics everywhere, but not a single word can we understand! Actually, understanding statistics is a critically important skill that we all need to have in this day and age. Every day, we are inundated with data about politics, sports, business, the stock market, health issues, financial matters, and many other topics. Most of us don't pay much attention to most of the statistics we hear, but more importantly, most of us don't really understand how to make sense of the numbers, ratios, and percentages with which we are constantly barraged. In order to obtain the truth behind the numbers, we must be able to ascertain what the data is really saying to us. We need to determine whether the data is biased in a particular direction or whether the true, balanced picture is correctly represented in the numbers. That is the reason for reading this book.

Statistics, as a field, is usually not the most popular topic or course in school. In fact, many people will go to great lengths to avoid having to take a statistics course. Often people think of it as a math course or something that is very quantitative, and that scares them away. Others, who get past the math, do not have the patience to search for what the numbers are actually saying. And still others don't believe that statistics can ever be used in a legitimate manner to point to the truth. But whether it is about significant trends in the population, average salary and unemployment rates, or similarities and differences across stock prices, statistics are an extremely important input to many decisions that we face daily. And understanding how to generate the statistics and interpret them relating to your particular decision can make all the difference between a good decision and a poor one.

For example, suppose that you are trying to sell your house and you need to set a selling price for it. The mean selling price of houses in your area is $250,000, so you set your price at $265,000. Perhaps $250,000 is the price roughly in the middle of several house prices that have ranged from $200,000 to $270,000, so you are in the ballpark. However, a mean of $250,000 could also occur with house prices of $175,000; $150,000; $145,000; $100,000; and $780,000. One high price out of five causes the mean to increase dramatically, so you have potentially priced yourself out of most of the market. For this reason, it is important to understand what the term "mean" really represents.

Another compelling reason to understand statistics is that we are living in a quality- and data-driven society. Everything nowadays is related to "improving quality," a "quality job," or "quality improvement processes" and there is now a lot more data available to consider. Companies are striving for higher quality in their products and employees and are using such programs as "continuous improvement" and "six-sigma" to achieve and measure this quality. Also now firms are using "big data" to make connections between what consumers purchase and a whole host of other variables that some would have previously considered to be irrelevant. So even the ordinary consumer needs to understand how these variables are linked together through statistics. That way they can make wiser purchase decisions.

As we move from the information age to the knowledge age, it is becoming increasingly important for us to at least understand, if not generate and use, statistics. In this book, Bob Donnelly and Fatma Abdel-Raouf have done a wonderful job of presenting statistics so that you can improve your ability to look at and comprehend the data you run across every day. Bob's and Fatma's many years of teaching statistics at all levels have provided the basis for their phenomenal ability to explain difficult statistical concepts clearly. Even the most unsophisticated reader will soon understand the subtleties and power of telling the truth with statistics!

Christine T. Kydd
Associate Dean for Undergraduate Programs
Professor and Area Head, Operations Management
Department of Business Administration
Lerner College of Business & Economics
University of Delaware

Introduction

Statistics. Why does this single word terrify so many of today's students? The mere mention of this word in the classroom causes a glassy-eyed, deer-in-the-headlights reaction across a sea of faces. In one form or another, the topic of statistics has been torturing innocent students for hundreds of years. You would think the word statistics had been derived from the Latin words sta, meaning "Why" and tistics, meaning "Do I have to take this %#!$@*% class?" But it really doesn't have to be this way. The term "stat" needn't be a four-letter word in the minds of our students.

As you read this paragraph, you're probably wondering what this book can do for you. Well, it's written by people (that's us) who (a) clearly remember being in your shoes as a student (even if it was in the previous century), (b) sympathize with your current dilemma (we can feel your pain), and (c) have learned a thing or two over many years of teaching (those many hours of tutorials were not for naught). The result of this experience has allowed us to discover ways to walk you through many of the concepts that traditionally frustrate students. Armed with the tools that you will gain from the many examples and numerous problems explained in detail, this task will not be as daunting as it first appears.

Unfortunately, fancy terms such as inferential statistics, analysis of variance, and hypothesis testing are enough to send many running for the hills. Our goal has been to show that these complicated terms are really used to describe ordinary, straight-forward concepts. By applying many of the techniques to everyday (and sometimes humorous) examples, we have attempted to show that not only is statistics a topic that anyone can master, but it can also actually make sense and be helpful in numerous situations.

So hold on to your hats, we're about to take a wild ride into the realm of numbers, inequalities, and, oh yes, don't forget all those Greek symbols! You will see equations that look like the Chinese alphabet at first glance, but can, in fact, be simplified into plain English. The step-by-step description of each problem will help you break down the process into manageable pieces. As you work the example problems on your own, you will gain confidence and success in your abilities to put numbers to work to provide usable information. And, guess what, that is sometimes how statisticians are born!

How This Book Is Organized

The book is organized into three parts:

In **Part 1, The Basics,** we start from the very beginning without any assumptions of prior knowledge. In this part, we dive into the world of data and learn about the different types of data and the variety of measurement scales that we can use. We also cover how to display data graphically, both manually and with the help of Microsoft Excel. We wrap up Part 1 with learning how to calculate descriptive statistics, such as the mean and standard deviation.

In **Part 2, Probability Topics,** we introduce the scary world of probability theory. Once again, we assume you have no prior knowledge of this topic (or if you did, we assume you buried it in the deep recesses of your brain, hoping to never uncover it). An important topic in this part is learning how to count the number of events, which can really improve your poker skills. After easing you into the basics, we gently slide into probability distributions, such as the normal and binomial. Once you master these, we have set the stage for Part 3.

In **Part 3, Inferential Statistics,** we start off learning about sampling procedures and the way samples behave statistically. When these concepts are understood, we start acting like real statisticians by making estimates of populations using confidence intervals. By this time, your own mother wouldn't recognize you! We'll top Part 3 off with a procedure that's near and dear to every statistician's heart—hypothesis testing. With this tool, you can do things like make bold comparisons between the male and female population. We'll leave that one to you.

Extras

Throughout this book, you will come across various sidebars that provide a helping hand when things seem to get a little tough. Many are based on our experience as teachers with the concepts that we have found to cause students the most difficulty.

DEFINITION

These are definitions of statistical jargon explained in a nonthreatening manner, which will help to clarify important concepts. You'll find that their bark is often far worse than their bite.

RANDOM THOUGHTS

In these sidebars we will give you insights that we find interesting (and hopefully you will, too!) about the current topic. Statistics is full of little-known facts that can help relieve the intensity of the topic at hand.

BOB'S BASICS

These are tips and insights that we have accumulated over the years of helping students master a particular topic. The goal here is to have that light bulb in the brain go off, resulting in the feeling of "I got it!"

WRONG NUMBER

These are warnings of potential pitfalls lying in wait for an unsuspecting student to fall into. By taking note of these, you'll avoid the same traps that have ensnarled many of your predecessors.

TEST YOUR KNOWLEDGE

In these sidebars we will give you an interesting application to the concept that we just covered so you can relate it to real world applications. We will also ask you if you see any connection between the topic covered and the application given in this sidebar. It is actually *fun!*

Acknowledgments

There are many people whom I am indebted to for helping me with this project. I'd like to thank Jessica Faust for her guidance and expertise to get me on track in the beginning, Mike Sanders for going easy on me with his initial feedback, and Nancy Lewis, for her valuable opinions during the writing process. I'd also like to thank Mike Thomas and Nancy Wagner for their helpful suggestions with the second edition.

To my colleague and friend, Dr. Patricia Buhler, who introduced me to the publishing industry, convinced me to take on this project, and encouraged me throughout the writing process. This all started with you, Pat.

To my in-laws, Lindsay and Marge, who never failed to ask me what chapter I was writing, which motivated me to stay on schedule. Your commitment to each other is a true inspiration for all of us.

To my boss of 10 years at Goldey-Beacom College, Joyce Jones, who rearranged my teaching schedule to accommodate my deadlines. Life at GBC will never be the same after you retire, Joyce. I am really going to miss you. Thank you for your constant support over the years. You have been a great boss and a true friend.

To my students who make teaching a pleasure. The lessons that I have leaned over the years about teaching were invaluable to me as I wrote this book. Without all of you, I would never have had the opportunity to be an author.

To my children, Christin, Brian, and John, and my stepchildren, Katie, Sam, and Jeff, for your interest in this book and your willingness to let me use your antics as examples in many of the chapters.

And most importantly, to my wife, Debbie, who made this a team effort with all the hours she spent contributing ideas, proofreading manuscripts, editing figures, and giving up family time to help me stay on schedule. Deb's excitement over my opportunity to write this book gave me the courage to accept this challenge. Deb was also the inspiration for many of the examples used in the book, allowing me to share experiences from our wonderful life together. Thank you for your love and your patience with me while writing this book.

—Bob Donnelly Jr., PhD

I would like to thank Jan Lynn, who helped me patiently throughout this process, never hesitated to provide me with advice, and answered my numerous questions. I also would like to thank Kayla Dugger, Alexandra Elliott, and Jerrell Cassady for their careful review and edit of the manuscript. Furthermore, I would like to thank Jessica Faust for all her help with this process and for arranging everything for me and connecting me with the editors I needed to work with.

To my colleague and friend, the late Dr. Robert Donnelly Jr., his wife Debbie, and their children. Bob passed away in June 2014. His sudden death was a shock to all of us. He left a hole that can't be filled. His friendly personality, his sense of humor, and his knowledge and extensive experiences are just a few of his traits that we all miss immensely.

To my family: my parents, my husband, my brothers, my mother-in-law, my sisters- and brothers-in-law, my nieces and nephew, my nieces- and nephew-in-law, my cousins, my uncles and aunts, and all my family, for your support throughout this process and for being there for me all the time. I'm blessed to have every one of you in my life. You are such a great and supportive family, and I could not have asked for anything more.

To my colleague and dear friend, Dr. Patricia Buhler, who recommended me to this project, and for her continuous help, support, and advice throughout the years. I'm blessed to have you as a friend, Pat!

To the academic dean at Goldey-Beacom College, Dean Alison White, for her accommodation to my teaching schedule and for going the extra mile to help us. To Dr. Gary Wirt, the President of Goldey-Beacom College for his support and accommodation. Dr. Wirt and Dean White have their doors open for us all the time, listening to our concerns and addressing them as fast as they can.

To my research assistant, Mary Elizabeth Rivers, who carefully reviewed the manuscript. Mary ensured that all equations and calculations were correct, she proofread the manuscript, and she reviewed the graphs and provided useful ideas.

To my students, who made teaching a pleasant experience, who enriched the classroom with their valuable comments and questions, and who made me an experienced teacher.

And above all, to the most important people in my life, my shining stars, whose love and support, patience and understanding, and passion and smiles I could not live without. Your existence in my life gives me the strength and the support to handle anything that comes my way. You are the pillars of my life and my life is meaningless without you! My shining stars are the best children any parent could ever dream of and I'm very blessed to have them. Loving you and looking at you gives me the strength to write this book. I could not have done it without you, my shining stars. I love you more than the whole world.

—Fatma Abdel-Raouf, PhD

Special Thanks to the Technical Reviewer

Idiot's Guides: Statistics, Third Edition, was reviewed by an expert who double-checked the accuracy of what you'll learn here, to help us ensure this book gives you everything you need to know about statistics. Special thanks are extended to Jerrell Cassady, PhD. Jerrell is a Professor of Psychology in the Department of Educational Psychology at Ball State University. He is also the Director of Graduate Programs in Educational Psychology and the Co-Director of the Research Design Studio. Jerrell was assisted by Christopher Thomas, MS, a doctoral candidate in the Department of Educational Psychology and a research fellow in the Research Design Studio.

Trademarks

PART

1

The Basics

The key to successfully mastering statistics is to have a solid foundation of the basics. To get a firm grasp of the more advanced topics, you need to be well grounded in the concepts presented in this part. The chapters in this part focus on data, the starting point for any method in statistics. You might be surprised with how much there really is to learn about data and all of its properties. We will examine the different types of data, how it is collected, how it is displayed, and how it is used to calculate things called the mean and standard deviation.

CHAPTER

1

Let's Get Started

In This Chapter

- The purpose of statistics—what's in it for you?
- The types of statistics—the major league and the minor league
- The ethical side of statistics

How many times have you asked yourself why you even need to learn statistics? Well, you're not alone. All too often students find themselves drowning in a mathematical swamp of theories and concepts and never get a chance to see the "big picture" before going under. Our goal in this chapter is to provide you with that broader perspective and convince you that statistics is a very useful tool in our current society. In other words, here comes your life preserver. Grab on!

In today's technologically advanced world, we are surrounded by a barrage of data and information from sources trying to convince us to buy something or simply persuade us to agree with their point of view. When we hear on TV that a politician is leading in the polls and in small print see + or − 4 percent margin of error, do we know what that means? When a new product is recommended by 4 out of 5 doctors, do we question the validity of the claim? (For instance, were the doctors paid for their endorsement? Or how many doctors were interviewed?) Statistics can have a powerful influence on our feelings, our opinions, and our decisions that we make in life. Getting a handle on this widely used tool is a good thing for all of us.

What Is Statistics and Why Do You Need It?

Statistics and data are all around us. I'm sure that not a single day passes without hearing lots of statistics. They are used to inform us and to influence us—to buy a specific product, to vote for a specific person, to sway our opinion about an issue, and much more. Examples of statistics that we hear are:

- In 2015, 114.4 million people watched the Super Bowl between the New England Patriots and the Seattle Seahawks. (Do you remember who won? The Patriots!)
- An August 2015 employment report showed that employers added 173,000 jobs in August and the unemployment rate decreased to 5.1 percent.
- The cost of a 30-second TV advertisement during the 2015 Super Bowl was $4.5 million. The cost of an advertisement during the first Super Bowl in 1967 was $42,000!
- In September 2015, a Gallup poll showed that President Obama's approval rate in the United States was 49 percent with a ± 3 percent margin of error. After reading this book, you will know exactly what that means!

So what is statistics? Statistics is the field of study that addresses collecting, summarizing, organizing, presenting, analyzing, and interpreting data. That's a mouthful! In simpler terms, statistics is a way to convert numbers into useful information so that you can make sound and informed decisions.

These decisions affect our lives in many ways. For instance, scientists conduct countless medical studies to determine the effectiveness of new drugs. Statistics form the basis of an objective decision as to whether a new drug is actually better or safer than current treatments.

Today's corporations make major business decisions based on statistical analysis. In the 1980s, Marriott conducted an extensive survey with potential customers on their attitudes about the company's different hotel options. After analyzing the data, the corporation launched Courtyard by Marriott, which has been a huge success.

The results of statistical studies and how they're presented also often dictate government policies. For example, the U.S. federal government heavily relies on the national census that is conducted every 10 years to determine funding levels for all the various parts of the country. The statistical analysis performed on this census data has far-reaching implications for ongoing programs at the state and federal levels.

The entire sports industry is completely dependent on the field of statistics. Can you even imagine the culture of baseball, football, or basketball without statistical analysis? You would never know who the top players are, who is currently hot, and who is in a slump. Without statistics, players could never negotiate those staggering salaries. Evidently we're onto something here, as statistics seems to infiltrate every part of our daily lives.

Now, why do you need to learn statistics? You need to pass the course in order to graduate and earn your degree. This is a common reason, but we're hoping that by the time you finish this book you will realize how important statistics is in your professional and personal life. No matter what your field of study or profession, chances are that statistics can help. As you go through this book, you will learn statistical techniques that enable you to read and understand the different statistical studies done in your field. You will also be able to perform your own studies. (Can't wait to be there, right? Keep reading!)

In addition, statistics can help you to make some of your personal choices in life as you become an informed and knowledgeable consumer of information and goods. With statistics, you will be able to better understand the results of studies you hear about different products, and you will make wiser purchasing decisions. Furthermore, you will be able to understand the results of different polls you hear in the news so that you are a smarter voter and citizen.

Our point here is to make you aware that statistics surround us in our society and that our world would be very different if this wasn't the case. Statistics is a useful, and sometimes even critical, tool in our everyday life.

Types of Statistics

Statistics has evolved into two main types: descriptive and inferential. *Descriptive statistics* is concerned with organizing, summarizing, and presenting data in a more meaningful way so that a reader can get the information he or she needs easily and quickly. *Inferential statistics*, on the other hand, enables the reader to make conclusions and inferences (that is where the name comes from) about the population based on a sample of the population. Because descriptive statistics is generally simpler, it can be thought of as the "minor league" of the field; whereas inferential statistics, being more challenging, can be considered the "major league" of the two.

The purpose of **descriptive** statistics is to summarize and present data in a more meaningful way so we can quickly obtain an overview of the data. **Inferential** statistics allows us to make claims or conclusions about a population based on a sample from that population.

Descriptive Statistics—the Minor League

Descriptive statistics plays an important role today because of the vast amount of data readily available at our fingertips. With a computer and an Internet connection, we can access volumes of data in seconds. Being able to accurately summarize all of this data to get a look at the "big picture," either graphically or numerically, is the job of descriptive statistics.

There are many examples of descriptive statistics. One of the commonly used descriptive statistics is the average. Let's say you are a marketing manager and you want to report to your supervisor the company monthly sales over the last five years. Imagine you could give your supervisor a long sheet of paper with 60 (5 years times 12 months) sales numbers on it! Or instead you could say that the average monthly sales over the last five years is $16,000. This way, your supervisor has an idea about how the company is doing quickly and easily, instead of having to review 60 numbers to find the same conclusion.

Descriptive statistics can also involve tables and graphs. In our example, you can report your average monthly sales per year in a table or a graph to show your supervisor that sales are increasing over the years. We will delve into descriptive statistics in more detail in Chapters 3 and 4. But until then, you are ready to move up to the big leagues—inferential statistics.

Inferential Statistics—the Major League

As important as descriptive statistics is to us number crunchers, we really get excited about inferential statistics. This category covers a large variety of techniques, which allow us to make actual claims about a population based on a sample of data from that population.

Suppose that I want to estimate the average weight of a bag of Herr's potato chips and see if it is close to my company's target weight of 1 ounce. Under-filling the potato chip bags will make customers unhappy, whereas overfilling the bags will make customers happy but will negatively affect my company's profit. To get the average weight of a bag of potato chips, I'll choose a sample of bags, weigh them, and get the average weight. Let's say the average is 0.9 ounces. Using inferential statistics, I might be able to determine the probability that the average weight of a bag of Herr's potato chips in the population (all bags of potato chips produced by Herr's) is actually the target 1 ounce even though the sample average is 0.9 ounces. Or I might conclude that the average weight of a bag of potato chips is less than 1 ounce, and I can go and check Herr's filling process and see why it is less than what it should be.

The following are more examples of inferential statistics:

- Based on a recent sample, I am 95 percent certain that the average age of my customers is between 32 and 35 years old.
- The average salary for male employees in a particular job category across the country was higher than the female employees' salary, based on a random survey.

In each case, the findings were based on a sample from a larger population and were used to make an inference on that entire population.

The basic difference between descriptive and inferential statistics is that descriptive statistics reports only on the observations at hand and nothing more. Inferential statistics makes a statement about a population based solely on results from a sample taken from that population.

We must tell you at this point that inferential statistics is the area of this field that students find the most challenging. To be able to make statements based on samples, you need to use mathematical models that involve probability theory. Now don't panic, take a deep breath, and count to 10 slowly. We realize that this is often the stumbling block for many, so we have devoted plenty of pages to that nasty "p" word.

Population vs. Sample

You will hear these two words a lot, so what do they mean? Population is the entire set of observations that the researcher is interested in, whereas a sample is a smaller portion of the population. A sample should be a scientifically chosen part of the population. It should be chosen in specific ways to make sure it is a representation of the population. Why do we use samples? Because in many cases, the population of interest is too large or too costly to collect data from so we use samples instead.

Figure 1.1 shows the relationship between an entire population and a sample.

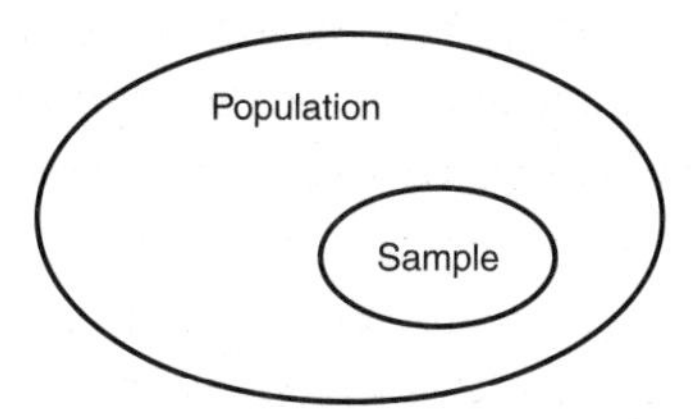

Figure 1.1
The relationship between a population and a sample.

Here is an example to illustrate the relationship between a population and a sample: On the first Friday of every month at 8:30 A.M., the employment report is released by the Bureau of Labor Statistics (BLS) and includes data about the number of jobs added in the previous month and the unemployment rate. To collect data and calculate these numbers, employees from the BLS interview a sample of the population and a sample of businesses. Now, can you imagine the employees of the BLS interviewing the entire U.S. population (over 320 billion people) and all of U.S. businesses to collect the data? If they did, it would take years to compose the employment report, and by the time it was released it would be too outdated to reflect accurately the current labor market. Therefore, the BLS just collects data from a sample of 60,000 households and 143,000 businesses and government agencies. Wise decision, right?

Parameter vs. Statistic

As you will learn, statistics people like to use all sorts of jargon, so here are a couple of terms. A *parameter* refers to data that is used to describe something of interest in the entire population, such as the average. However, if the data describes a sample from that population, it's called a *statistic*. For example, let's say the population of interest is Bob's wife Debbie's three-year-old preschool class and our measurement of interest is how many times the children use the bathroom in a day (according to Debbie, much more than should be physically possible!).

If we average the number of trips per child, this figure would be considered a parameter because the entire population was measured. However, if we want to make a statement about the average number of bathroom trips per day per three-year-old in the country, then Debbie's class would be a sample. We can consider the average that we observe from her class a statistic if we assume it could be used to estimate all three-year-olds in the country.

Another example is grades for students: Imagine I have a large statistics class of 200 students, and I want to find the average grade for the class. If I use the grades for all 200 students, then the average I get is called a parameter. If I use a sample of say 50 students, then the average I get is called a statistic, which I use as an estimate of the parameter.

DEFINITION

Parameter is a descriptive measure of the population. **Statistic** is a descriptive measure of the sample, and we use it as an estimate of the parameter.

Try to get used to saying population parameter and sample statistic. To make it easy for you to remember, both population and parameter start with a "p," and both sample and statistic start with an "s."

TEST YOUR KNOWLEDGE

What is the difference between statistics (with an "s") and statistic (without an "s")? **Statistics** is a field of study that addresses collecting, organizing, summarizing, presenting, analyzing, and interpreting data. A **statistic** is a descriptive measure of the sample, which we use as an estimate of the population parameter.

Ethics and Statistics—It's a Dangerous World Out There

People often use statistics when attempting to persuade you to their point of view. Because they are motivated to convince you to purchase something from them or simply to support them, this motivation can lead to the misuse of statistics in several ways.

One of the most common misuses is choosing a sample that ensures results consistent with the desired outcome, rather than choosing a sample representative of the population of interest. This is known as having a *biased sample.*

Suppose that we're lobbying for the golf industry and want to convince Congress to establish a national golf holiday. During this honored day, all government and business offices would be closed so that we could all run out to chase a little white ball into a hole that's way too small, with sticks purposely designed by the evil golf companies to make this task impossible. Sounds like fun to us! Somehow, we would need to demonstrate that the average level-headed American is in favor of this. Here is where the genius part of our plan lies: rather than survey the general American public, we pass out the survey form only at golf courses. But wait ... it only gets better. We design the survey to look like the following:

> We would like to propose a national golf holiday, on which everybody gets the day off from work and plays golf *all day.* (This means you would not need permission from your spouse.) Are you in favor of this proposal?
>
> A. Yes, most definitely.
>
> B. Sure, why not?
>
> C. No, I would rather spend the entire day at work.
>
> P.S. If you choose C, we will permanently revoke all your golfing privileges everywhere in the country for the rest of your life. We are dead serious.

We can now honestly report back to Congress that the respondents of our survey were overwhelmingly in favor of this new holiday.

This might be an extreme example, but it demonstrates that we can easily manipulate survey results by the way that we ask questions and by whom we choose to survey.

Another way to misuse statistics is through misleading graphs that make differences seem larger or smaller depending on the way the data are graphed. For example, the following data represents the unemployment rate in the United States from 2000 to 2008.

January of Each Year	Unemployment Rate
2000	4.0
2001	4.2
2002	5.7
2003	5.8
2004	5.7
2005	5.2
2006	4.7
2007	4.6
2008	4.9

Now, look at Figures 1.2 and 1.3, which represent the same data.

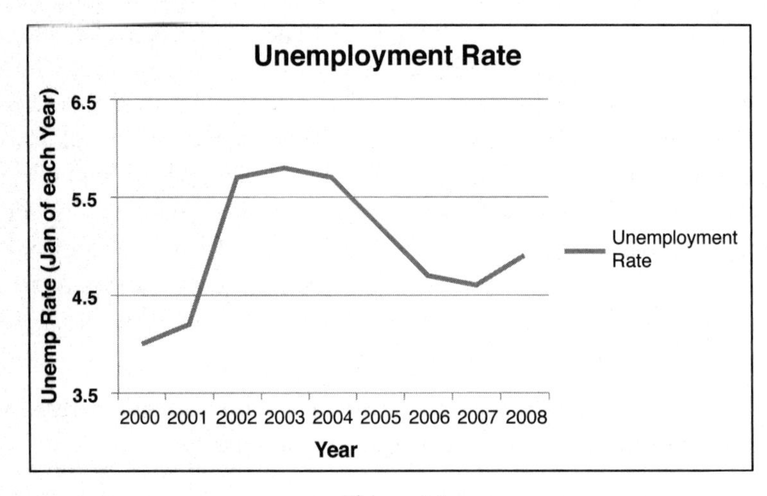

Figure 1.2

This graph shows the change in unemployment rate in the United States with a magnified scale on the axis.

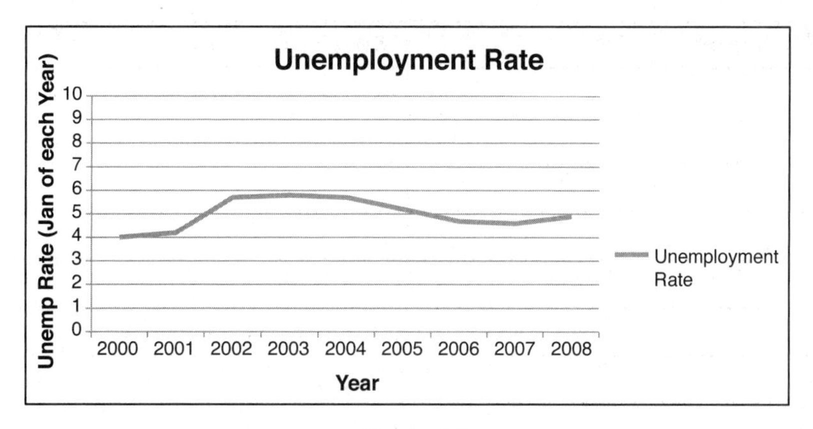

Figure 1.3

This graph shows a less exaggerated picture of the unemployment rate fluctuation in the United States.

If I want to give you the impression that there was a large increase in the unemployment rate in the United States from 2000 to 2008, I'll show you the graph in Figure 1.2, whereas if I want to show you that the unemployment rate barely increased from 2000 to 2008, I'll show you the graph in Figure 1.3. What is the difference between the two graphs? Even though they represent the same data and should give you the same impression no matter how I graph it, I used a different scale on the vertical axis for each one. In Figure 1.2, the vertical axis shows from 3.5 percent to 6.5 percent only, so the increase in the unemployment rate, while minimal in the data, shows as magnified in the graph. In Figure 1.3, I used a regular scale from 0 percent to 10 percent, so the increase in the unemployment rate reflects the slightness of the changes. So by changing the scale on the graph, I can choose how exaggerated, and perhaps biased, the overall picture of the data is.

There are many other examples of misleading graphs and representations of data and statistics, so always be careful when looking at any graph or reading any statistic.

A third way for statistics misuse is Internet surveys and polls. Many websites encourage visitors to vote on a question of the day. The results of these informal polls are unreliable simply because those collecting the data have no control over who responds or how many times a person responds. As stated earlier, a reliable statistical study depends on selecting a sample representative from the population of interest. This is not possible when any person surfing the Internet can participate in the poll. Even though most of these polls state that the results are not scientific, it's still a natural human tendency to be influenced by the results we see.

The lesson here is that we are all consumers of statistics. We are constantly surrounded by information provided by someone who is trying to influence us or gain our support. By having a basic understanding about the field of statistics, we increase the likelihood that we can ward off those evil spirits in their attempts to distort the truth. In Chapter 2, we'll begin our journey to achieve this goal…oh, and to help you pass your statistics course.

Practice Problems

Identify each of the following statistics as either descriptive or inferential.

1. Seventy-three percent of Asian American households in the United States own a computer.
2. Households with children under the age of 18 are more likely to have access to the internet (62 percent) than family households with no children (53 percent).
3. Hank Aaron hit 755 career home runs.
4. The average SAT score for incoming freshman at a local college was 950.
5. On a recent poll in January 2013, President Obama's approval rate was 52 percent in the United States.

The Least You Need to Know

- Statistics is a vital tool that provides people and organizations with the necessary information to make good decisions.
- Descriptive statistics focuses on the summary or display of data so we can quickly obtain an overview.
- Inferential statistics allows us to make claims or conclusions about a population based on a sample of data from that population.
- We are all consumers of statistics and need to be aware of the potential misuses that can occur in this field.

CHAPTER

2

Data, Data Everywhere and Not a Drop to Drink

Data is the basic foundation for the field of statistics. The validity of any statistical study hinges on the validity of the data from the beginning of the process. Many things can come into question, such as the accuracy of the data or the source of the data. Without the proper foundation, your efforts to provide a sound analysis will come tumbling down.

The issues surrounding data can be surprisingly complex. After all, aren't we just talking about numbers here? What could go wrong? Well, plenty can. Because data can be classified in different ways, we need to recognize the difference between them. Data also can be measured in many ways. The data measurement choice we make at the start of the study will determine what kind of statistical techniques we can apply.

In This Chapter

- The difference between data and information
- Where does data come from?
- What kinds of data and variables can we use?
- Classifying data by their level of measurement
- Setting up Excel for statistical analysis

The Importance of Data

Data is simply defined as the value assigned to a specific observation or measurement. If Bob is collecting data on his wife's snoring behavior, he can do so in different ways. He can measure how many times Debbie snores over a 10-minute period. He can measure the length of each snore in seconds. He could also measure how loud each snore is with a descriptive phrase, like "That one sounded like a bear just waking up from hibernation!" or "Wow! That one sounded like an Alaskan seal calling for its young." (How a sound like that can come from a person who can fit into a pair of size 2 jeans and still be able to breathe, you'll never know.)

In each case, we're recording data on the same event in a different form. In the first case, we're measuring a frequency or number of occurrences. In the second instance, we're measuring duration or length in time. And the final attempt measures the event by describing volume using words rather than numbers. Each of these cases just shows a different way to use data.

Data is the building blocks of all statistical studies. You can hire the most expensive, well-known statisticians and provide them with the latest computer hardware and software available, but if the data you provide them is inaccurate or not relevant to the study, then the final results will be worthless.

However, data all by its lonesome is not all that useful. By definition, data is just the raw facts and figures that pertain to a measurement of interest. *Information*, on the other hand, is derived from the facts for the purpose of making decisions. One of the major reasons to use statistics is to transform data into information. For example, the table that follows shows monthly sales data for a small retail store.

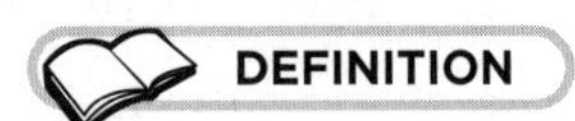

Data is the value assigned to an observation or a measurement and is the building blocks to statistical analysis. **Information,** on the other hand, is data that is transformed into useful facts that can be used for a specific purpose, such as making a decision.

Monthly Sales Data

Month	Sales ($)
January	15,178
February	14,293
March	13,492
April	12,287
May	11,321

Using statistical analysis, we can generate information that may be of interest, such as "Wake up! You are doing something very wrong. At this rate, you will be out of business by early next year." Based on this valuable information, we can make some important decisions about how to avoid this impending disaster.

TEST YOUR KNOWLEDGE

Is the word "data" singular or plural? In the dictionary, "data" is a plural noun. The singular is "datum." However, people commonly now use the word "data" as both plural and singular, depending on what it references. If "data" is used as a count noun (nouns that can be counted; describe "how many"), then it is plural. For example, data include names, ages, gender, and race of students. If "data" is used as a mass noun (nouns that cannot be counted; describe "how much"), then people often use it as singular. For example, data is available everywhere.

The Sources of Data—Where Does All This Stuff Come From?

We classify the sources of data into two broad categories: *primary* and *secondary*. Secondary data is data that somebody else has collected and made available for others to use. The U.S. government loves to collect and publish all sorts of interesting data, just in case anyone should need it. The Department of Commerce handles census data, and the Department of Labor collects mountains of, you guessed it, labor statistics. The Department of the Interior provides all sorts of data about U.S. resources.

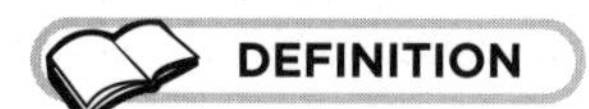

DEFINITION

Primary data is data that you have collected for your own use. **Secondary** data is data collected by someone else that you are "borrowing."

The main drawback of using secondary data is that you have no control over how the data was collected. It's a natural human tendency to believe anything that's in print (you believe us, don't you?), and sometimes that requires a leap of faith. The advantage of secondary data is that it's cheap (sometimes free) and it's available immediately. That's called instant gratification.

Primary data, on the other hand, is data collected by the person who eventually uses this data. It can be expensive to acquire, but the main advantage is that it's your data and you have control over how it's gathered. Then you have nobody else to blame but yourself if you make a mess of it.

When collecting primary data, you want to ensure that the results will not be biased by the manner in which they are collected. You can obtain primary data in many ways, such as direct observation, surveys, and experiments.

Direct Observation—I'll Be Watching You

Whether the subjects know it or not, direct observation involves observing behavior as it occurs. Most often, this method focuses on gathering data while the subjects of interest are in their natural environment, oblivious to being watched (called disguised observation). Examples of these studies would be observing wild animals stalking their prey in the forest or teenagers at the mall on Friday night (or is that the same example?). The advantage of this method is that the subjects will unlikely be influenced by the data collection.

Focus groups are a direct observational technique where the subjects are aware that data is being collected (undisguised observation). Businesses use focus groups to gather information in a group setting controlled by a moderator. The subjects are asked to discuss specific topics and are usually paid for their time.

Experiments—Who's in Control?

This method is more direct than observation because the subjects will participate in an experiment designed to determine the effectiveness of a treatment. An example of a treatment could be the use of a new medical drug. Two groups would be established. The first is the experimental group who receive the new drug, and the second is the control group who thinks they are getting the new drug but are in fact getting no medication. The reactions from each group are measured and compared to determine whether the new drug is effective.

The benefit of experiments is that they allow the statistician who designs the experiment to observe how certain manipulated variables could influence the results, such as gender, age, and education of the participants. The concern about collecting data through experiments is that the response of the subjects might be influenced by the fact that they are participating in a study. In addition, the claims that the experimental studies are attempting to verify need to be clear and specific. The design of experiments for a statistical study is a very complex topic and goes beyond the scope of this book.

Surveys—Is That Your Final Answer?

This technique of data collection involves asking the subject a series of questions. The questionnaire needs to be carefully designed to avoid any bias (see Chapter 1) or confusion for those participating. Concerns also exist about the influence that the survey will have on the participant's responses. Some participants tend to adjust their responses to fit in line with what they believe is socially desirable or "the right answer." The survey can be administered online, or by e-mail, snail-mail, or telephone. It's the telephone survey that I'm most fond of, especially when I get the call just as I'm sitting down to dinner, getting into the shower, or finally making some progress on the chapter I'm writing.

Whatever method you employ, your primary concern should always be that the sample is representative of the population in which you are interested.

BOB'S BASICS

Research has shown that the manner in which the questions are asked can affect the responses that a person provides on a questionnaire. A question posed in a positive tone will tend to evoke a more positive response and vice versa. A good strategy is to test your questionnaire with a small group of people before releasing it to the general public.

Data and Types of Variables

Variables are measures of characteristics of interest. The variable is the basic element of algebra, and it can take on any value (numerical or non-numerical). Data are the values that the variable takes. The range of values that each variable can have differs significantly.

Variables can be divided according to different criteria, such as:

- Quantitative versus qualitative
- Discrete versus continuous
- Dependent versus independent

Quantitative vs. Qualitative Variables

A quantitative variable is a variable with numerical values such as age and income. A qualitative variable is a variable with descriptive, non-numerical values, such as gender and race. If I ask you about your gender, you can describe yourself as male or female. The answer is a descriptive value, not a number, so it is a qualitative variable. Whereas if I ask you about your age, you will give me a number so it is a quantitative variable. If you ask me, I'd say I'm 21 years old!

Discrete vs. Continuous Variables

Quantitative variables can be classified as either discrete or continuous. A discrete variable assumes certain values only, usually with gaps between the values. It results from counting or enumeration. For example, the number of cars or children is a discrete value. If I ask you how many cars you have, you could say "one," "two," or "three." But can you say, "I have 1.5 cars?" Or can you say, "I have 2.5 kids?" Of course not! This is a discrete variable and its value is usually a whole digit with no decimal or fraction. A continuous variable, on the other hand, can take on any value within a certain range, such as length, distance, and height. It results from measurements. For example, you can say the distance from my home to the nearest mall is 5.6 miles.

Dependent vs. Independent Variables

A dependent variable is a variable whose value is determined by another variable. An independent variable is a variable whose value is unaffected by another variable. For example, your exam grade depends on how long you studied. Here, your exam grade is the dependent variable and your hours of study are the independent variables. The dependent variable responds to the independent variable. That's why it is called the dependent variable—because it *depends* on the independent variable.

Types of Measurement Scales—a Weighty Topic

An important way to classify data is by the way it is measured. This distinction is critical because it affects which statistical techniques we can use in our analysis of the data. Measurement classification can be made in several levels: nominal measurement, ordinal measurement, interval measurement, and ratio measurement. Each measure adds something to the previous one. Let's look at each.

Nominal Level of Measurement

A *nominal* level of measurement deals with qualitative variables. In nominal measurement, names or classifications are used to divide the data into separate categories, with no meaningful order. One example is gender, with the categories being male and female. I could assign the number 0 for males and 1 for females, or I could switch them to 1 for males and 0 for females. But as you can see in this example, the order or value of the number assigned to the category is unimportant. In other words, we cannot reasonably rank-order the numbers from highest to lowest because only the qualitative measurements are meaningful, not the numbers assigned to the categories.

Other examples of nominal data are zip codes (there is no meaningful order or ranking to zip codes), marital status, race, types of dogs, and so forth.

This measurement type does not allow us to perform any mathematical operations, such as adding or multiplying. These types of data are considered the lowest level of data. As a result, it is the most restrictive when choosing a statistical technique to use for analysis.

Ordinal Level of Measurement

On the food chain of data measurement, *ordinal* is the next level up. It has all the properties of nominal data with the added feature that we can rank-order the categories or values from lowest to highest. The ordinal level of measurement can be qualitative or quantitative data. In ordinal measurements, the order of the numbers or values is meaningful while the magnitude of the values is not.

An example of ordinal measurements would be a survey with the options: strongly disagree (1), disagree (2), neutral (3), agree (4), and strongly agree (5). Here, the order of the numbers is important since (4) is better than (2), but the magnitude of the number assigned to the category is not important. I cannot say that (4) is twice as good as (2). Another example of the ordinal level of measurement is movie ratings with 1, 2, 3, or 4 stars. We know a 4-star movie is better than a 1-star movie, however we cannot claim that a 4-star movie is 4 times as good as a 1-star movie.

Interval Level of Measurement

Moving up the scale of data, we find ourselves at the interval level, which deals with quantitative variables only. With interval measurements, both the order and the magnitude of the numbers are meaningful because the distance between the measurements is quantitatively equidistance. Now we get to work with the mathematical operations of addition and subtraction when comparing values. For this data, we can measure the difference between the categories with numbers that actually provide meaningful information. An example is temperature: 80°F is 10 degrees warmer than 70°F. However, ratio comparisons don't make sense, so we cannot perform multiplication and division on this type of data. Why not? Simply because we cannot argue that 80°F is twice as warm as 40°F.

Another characteristic of interval measurements is that there is no true zero point—zero does not mean that there is no quantity. For example 0°F or 0°C does not mean that there is no temperature (even though they feel very cold!). Other examples of interval data are IQ scores, SAT scores, and GPAs.

Ratio Level of Measurement

The king of measurement types is the *ratio* level, which are quantitative variables. This is as good as it gets as far as data is concerned. Now we can perform all four mathematical operations to compare values with absolutely no feelings of guilt. Ratio data has all the features of interval data (meaningful order and magnitude) with the added benefit of a true zero point. The term "true zero point" means that a 0 data value indicates the absence of the object being measured. An example of ratio level data is weight: 80 pounds are 10 pounds more than 70 pounds, 0 pounds actually indicates the absence of any pounds, and 80 pounds are twice as heavy as 40 pounds (we cannot say this for the interval measurement). Other examples include distance, age, height, time, and salary.

With a true 0 point, ratios make sense and we can use the rules of multiplication and division to compare data values. This allows us to say that a person who is 6 feet in height is twice as tall as a 3-foot person or that a 20-year-old person is half the age of a 40 year old.

The distinction between interval and ratio data is a fine line. To help identify the proper scale, use the "twice as much" rule. If the phrase "twice as much" accurately describes the relationship between two values that differ by a multiple of 2, then the data can be considered ratio level.

Interval data does not have a true 0 point. For example, 0 degrees Fahrenheit does not represent the absence of temperature, even though it may feel like it. To help explain this, try baking a cake at twice the recommended temperature in half the recommended time. Yuck!

Figure 2.1 summarizes the different data scales and how they relate to one another. As we explore different statistical techniques later in this book, we will revisit these different measurement scales. You will discover that specific techniques require certain types of data.

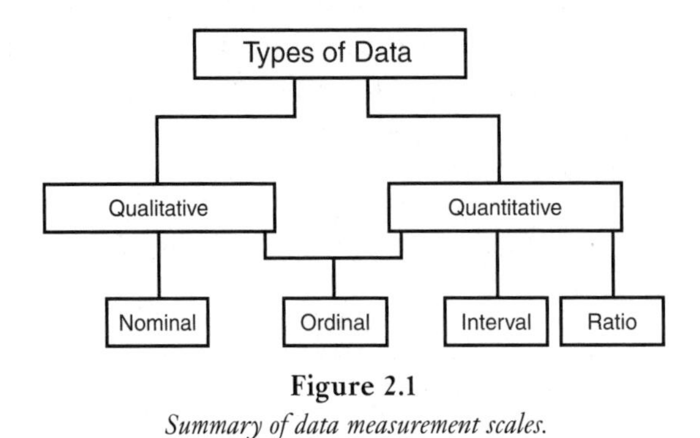

Figure 2.1
Summary of data measurement scales.

Computers to the Rescue

As mentioned in Chapter 1, we will explore the use of Excel in solving some of the statistics problems in this book. If you have no interest in using Excel in this manner, just skip this section. We promise you won't hurt our feelings. The purpose of this last section is to talk about the use of computers with statistics in general and then to make sure that your computer is ready to follow along.

The Role of Computers in Statistics

During the 1970s and 1980s, people with high levels of programming skill were performing on mainframe computers the only serious statistical analysis. These people were somewhat "different" from the rest of us. Fortunately, we have advanced from the Dark Ages and now have awesome, user-friendly computing power at our fingertips. Powerful programs such as SAS, SPSS, Stata, Minitab, R, and Excel are readily accessible to those of us who don't know a lick of computer programming and allow us to perform some of the most sophisticated statistical analysis known to humankind.

Parts of this book will demonstrate how to solve some of the statistical techniques using Microsoft Excel. Choosing to skip these parts will not interfere with your grasp of topics in subsequent chapters. This is simply optional material to expose you to statistical analysis on the computer. I also assume you already have a basic working knowledge of how to use Excel.

Installing the Data Analysis Add-In

Our first task is to check whether Excel's data analysis tool is available on your computer. If you are a Mac user, unless you have Excel 2016 for Mac, then you will not be able to use this add-in without a third-party installation (so feel free to research your options!). For PC users, follow along:

- Open an Excel spreadsheet and click on the Data tab at the top of the Excel window. If you see the Data Analysis tab as in Figure 2.2 below, then you are all set and don't need to install anything. You can safely skip the rest of this chapter and move on to Chapter 3.

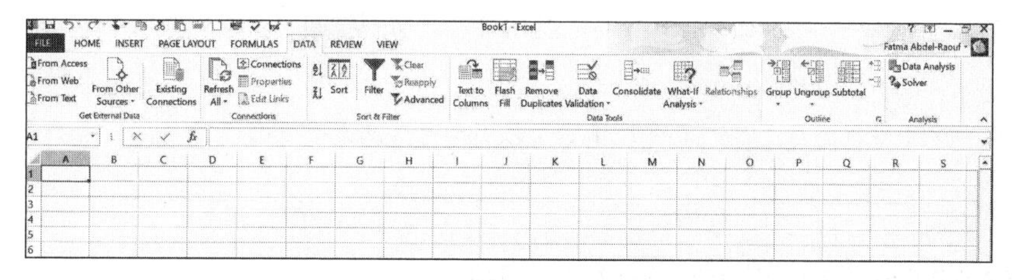

Figure 2.2
Excel's Data Analysis add-in.

- If you don't see the Data Analysis tab, such as in Figure 2.3, then you need to install it by following the steps below.

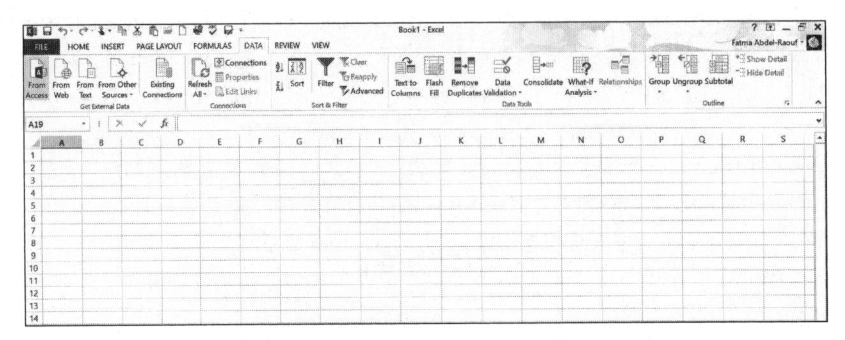

Figure 2.3
Excel without the Data Analysis add-in.

1. Click on File then choose Options. Select Add-Ins and click on Analysis ToolPak.

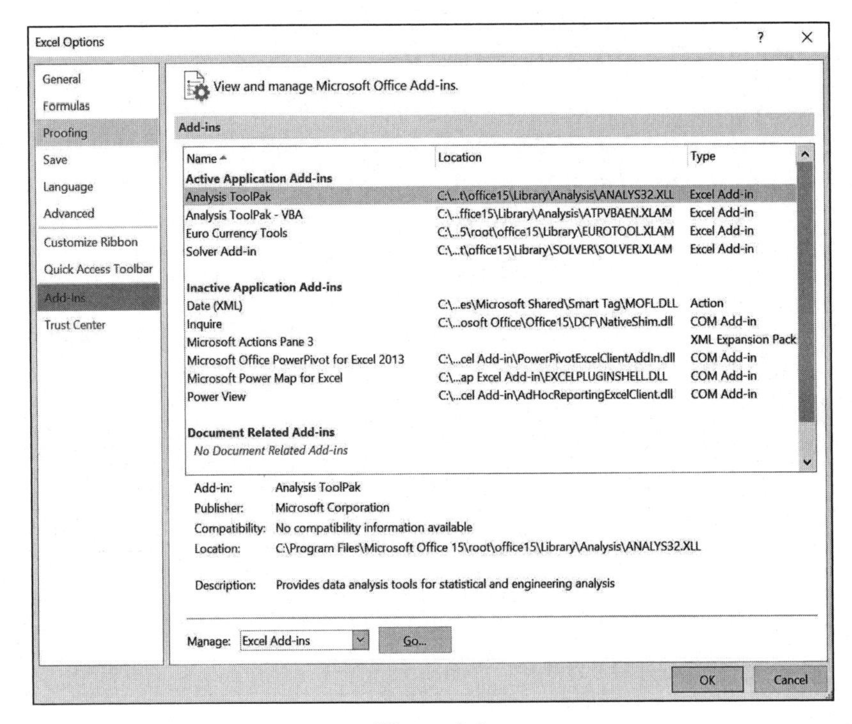

Figure 2.4
Adding Excel's Options.

2. Click on Go. And then check the boxes for Analysis ToolPak and Analysis ToolPak—VBA. Solver is also useful so you may want to check it, too. Then click OK.

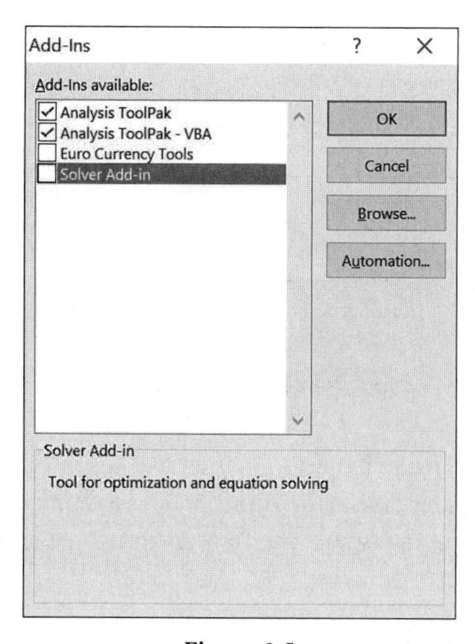

Figure 2.5
Excel's Add-Ins dialog box.

3. Click on the Data tab, and you will now see the Data Analysis icon displayed on the right hand side.

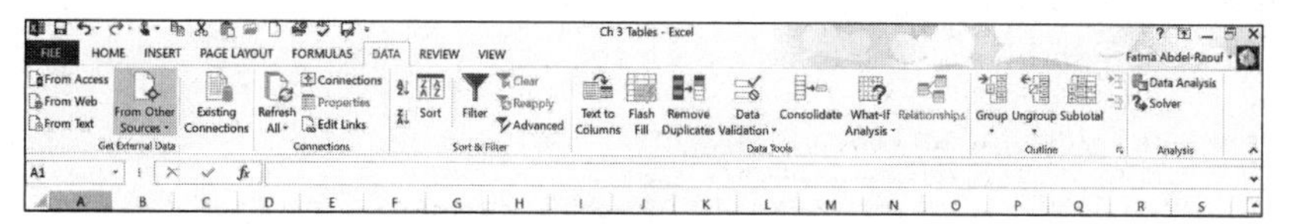

Figure 2.6
Excel with the Data Analysis Add-In.

4. Click on Data Analysis, and you will see the Data Analysis menu, shown in Figure 2.7.

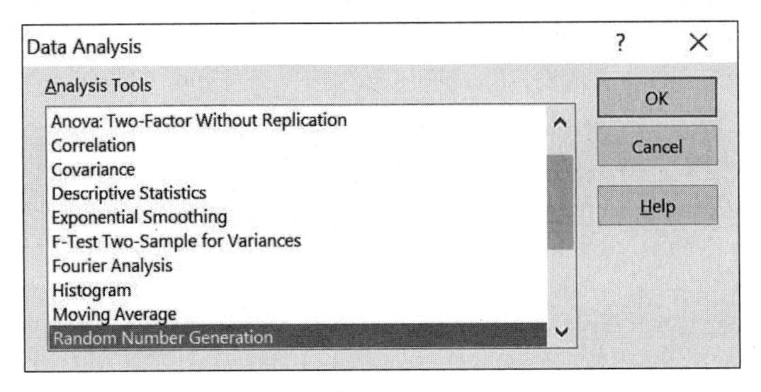

Figure 2.7
Excel's Data Analysis dialog box.

Your Excel program is now ready to perform all sorts of statistical magic for you as we explore various techniques throughout this book. At this point, you can click Cancel and close out Excel. Each time you open Excel in the future, the Data Analysis icon should be available under the Data tab at the top of the Excel window.

Practice Problems

Classify the following data as nominal, ordinal, interval, or ratio. Explain your choice.

1. Average monthly temperature in Fahrenheit degrees for the city of Wilmington throughout the year
2. Average monthly rainfall in inches for the city of Wilmington throughout the year
3. Education level of survey respondents

Level	Number of Respondents
High school	168
Bachelor's degree	784
Master's degree	212

4. Marital status of survey respondents

Status	Number of Respondents
Single	28
Married	189
Divorced	62

5. Age of the respondents in the survey
6. Gender of the respondents in the survey
7. The year in which the respondent was born
8. The voting affiliations of the respondents in the survey, classified as Republican, Democrat, or Undecided
9. The race of the respondents in the survey, classified as White, African American, Asian, or Other
10. The performance rating of employees, classified as Above Expectations, Meets Expectations, or Below Expectations
11. The uniform number of each member on a sports team
12. A list of the graduating high school seniors by class rank
13. The final exam scores for my statistics class on a scale of 0 to 100
14. The state in which the respondents in a survey reside

The Least You Need to Know

- Data serves as the building blocks for all statistical analysis.
- Data are the values that variables can take. Variables can be classified as quantitative or qualitative; discrete or continuous; and dependent or independent.
- Nominal data is assigned to categories with no mathematical comparisons between observations.
- Ordinal data has all the properties of nominal data with the additional capability of arranging the observations in order.
- Interval data has all the properties of ordinal data with the additional capability of calculating meaningful differences between the observations.
- Ratio data has all the properties of interval data with the additional capability of expressing one observation as a multiple of another.

CHAPTER

3

Displaying Descriptive Statistics

Having explained the various types of data that exist for statistical analysis in Chapter 2, here we will explore the different ways in which we can present data. In its basic form, making sense of the patterns in the data can be very difficult because our human brains are not very efficient at processing long lists of raw numbers. We do a much better job of absorbing data when it is presented in a summarized form through tables and graphs.

In the next several sections, we will examine many ways to present data so that it will be more useful to the person performing the analysis. Through these techniques, we are able to get a better overview of what the data is telling us. And believe me, there is plenty of data out there with some very interesting stories to tell. Stay tuned.

Because data is all around us, an important part of statistics is to know how to organize and present the data in a meaningful way so that the reader can get the information he or she needs quickly. As an example, Goldey-Beacom College (where we work) has over 2,000 students. The president of the college wants to know the percentage of students with a GPA of 3.00 or higher. GBC's President asks the admissions dean for the students' GPA data. The dean gave him a very long list

In This Chapter

- How to construct different frequency distributions
- How to graph a frequency distribution with a histogram
- The usefulness of pie, bar, and line charts
- How to construct a stem and leaf display
- Using Excel to construct charts

with the name and GPA of each student at GBC. How useful is this for the president to get the information he needs? How long will it take him to figure the percentage of students with a GPA of 3.00 or higher? Yes, I hear you, a really long time. Instead, the admissions dean can give him the same information organized in a table, such as the one below, and GBC's President can find the percentage of students with a GPA of 3.00 or higher in a second, literally!

Relative Frequency Distribution

GPA	Number of Students	Percentage of Students
1.00-1.50	50	3%
1.51-2.00	100	5%
2.01-2.50	300	15%
2.51-3.00	450	23%
3.01-3.50	600	30%
3.51-4.00	500	25%
Total	2000	100%

This table is an example of the relative frequency distribution that we will see in this chapter.

Organizing and presenting data sets can take three main forms: frequency distributions, graphs, and stem and leaf designs.

Frequency Distributions

Frequency distributions can take several forms:

- Frequency distribution
- Relative frequency distribution
- Cumulative frequency distribution
- Contingency table

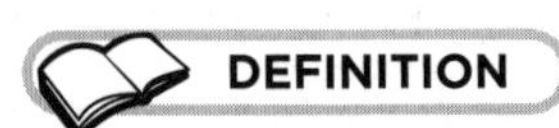

A **frequency distribution** is a table with two columns: One column has the classes for the variable of interest and the second column has the frequency in each class.

The Frequency Distribution

A frequency distribution is the basic table type. Once you master it, you can add another column for each of the next two table types to give the reader more information. To see how useful frequency distributions are, let's look at an example. The grades for 20 students in my statistics class are as follows:

93	81	92	75	78
82	65	98	62	84
58	64	73	85	59
56	72	87	91	71

You are my helper today, so I asked you to tell me the number of students with grades in the 80s and 90s. Doing so by looking at the data above will take you a long time. But you remembered your statistics class and said, "I'll use the information I learned in class." You know we can present the data in a frequency distribution by dividing the variable (the grades) into ranges and counting the frequency of the variables in each range.

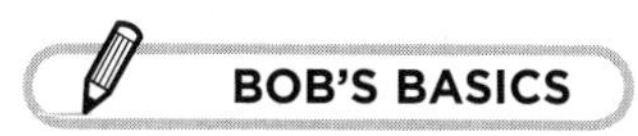

The intervals in the frequency distribution are known as *classes*, and the number of observations in each class is known as *frequency*.

To construct the frequency distribution, you need to do three things:

- Determine the number of classes
- Determine the class width
- Count the observations in each class

Doing these steps, you came up with the following frequency distribution:

Frequency Distribution

Grade Range	Number of Students
51 - 60	3
61 - 70	3
71 - 80	5

continues

Frequency Distribution (continued)

Grade Range	Number of Students
81 - 90	5
91 - 100	4
	Total = 20

I looked at the frequency distribution and said, "Well done!" In just a second, I can see that I have 5 students with grades in the 80s and 4 students with grades in the 90s. How long would have taken me to get this information by looking at the data themselves? And those are only 20 observations. Imagine if they were 2,000 observations!

In constructing the frequency distribution, make sure you follow these rules:

- Form classes of equal size. In the example above, we assigned 10 data values to be in each class. The first class includes grades of 51, 52, 53, 54, 55, 56, 57, 58, 59, and 60. 61 goes into the next class.
- Make classes mutually exclusive. Or in other words, avoid overlapping classes. In the example above, don't use 51 – 60 and 60 – 69. If you use overlapping classes, then a student with a grade of 60 would go into both of the different classes, and you would be double-counting the student.
- Avoid open-ended classes, if possible (for instance, the last class shouldn't be 91 and higher).
- Make sure you include all the data. Design the classes to include the lowest and the highest observations in your data—in other words, the classes should be exhaustive.

I know the question you want to ask me is, "how many classes should I have?" Try to have a reasonable number of classes for your data. Too few or too many classes will obscure patterns in a frequency distribution. Consider an extreme case where there is only one class with all the observations in it. The other extreme case is where we have too many classes and each class has only one observation. In this case, you didn't really organize the data, and it would be a pretty useless frequency distribution!

(A Distant) Relative Frequency Distribution

Now you did such a good job with the frequency distribution, so I asked you to do one more thing for me. I want to know the percentage of students with grades in the 80s and 90s. You think to yourself, "This is easy. I'll just add a column to the table for the percentage." You came up with the following table:

Relative Frequency Distribution

Grade Range	Number of Students	Percentage of Students
51 - 60	3	$\frac{3}{20}$ = 15%
61 - 70	3	$\frac{3}{20}$ = 15%
71 - 80	5	$\frac{5}{20}$ = 25%
81 - 90	5	$\frac{5}{20}$ = 25%
91 - 100	4	$\frac{4}{20}$ = 20%
	Total = 20	= 100%

This is a *relative frequency distribution.* Rather than to just display the number of observations in each class, the relative frequency distribution calculates the percentage of observations in each by dividing the frequency of each class by the total number of observations.

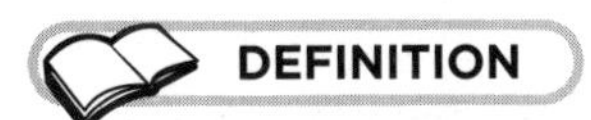

Relative frequency distributions display the percentage of observations in each class relative to the total number of observations. The percentages are called relative frequencies.

Another job well done! In a second, I can tell that 25 percent of students in my class got 80s, and 20 percent got 90s!

The Cumulative Frequency Distribution

Now I want to know the number of students in the class with grades less than or equal to 80. I can tell you are thinking that this is easy, and you just add the frequency in each class to the frequencies of all previous classes. In just a few minutes, you give me the following *cumulative frequency distribution*:

Cumulative Frequency Distribution

Classes	Frequency	Cumulative Frequency
51 - 60	3	3
61 - 70	3	3 + 3 = 6
71 - 80	5	5 + 3 + 3 = 11

continues

Cumulative Frequency Distribution (continued)

Classes	Frequency	Cumulative Frequency
81 - 90	5	5 + 5 + 3 + 3 = 16
91 - 100	4	4 + 5 + 5 + 3 + 3 = 20
	Total = 20	

I look at the table and say, "I've 16 students in the class with a grade of 90 or less and no student with a grade less than 50." This looks like a good class!

The **cumulative frequency distribution** displays the number of observations that are less than or equal to the current class. In other words, it sums the frequency in the current class and the frequencies in the previous classes.

What is the difference between cumulative frequency and relative frequency? Cumulative frequency is determined by *adding* the frequency in a class to all previous frequencies. Relative frequency is determined by *dividing* the frequency in a class by the total frequency.

The Contingency Table

You like the tables you created, and now you ask yourself, "What if I want to see the data for two variables instead of one?" You have both the grades and the genders of the students, and you want to see the distribution of grades by gender. So you asked me how you could do that, and I said, "This is what the contingency table is all about." The *contingency table* organizes data for two variables simultaneously. You have the following data for the 20 students:

93 (F)	81 (F)	92 (M)	75 (F)	78 (M)
82 (M)	65 (F)	98 (F)	62 (M)	84 (F)
58 (F)	64 (F)	73 (M)	85 (M)	59 (M)
56 (F)	72 (M)	87 (M)	91 (F)	71 (M)

You start by making the frequency distribution table, but with two frequency columns, one for each gender. You came up with the following contingency table:

Contingency Table

Classes	M	F	Total
51 - 60	1 (5%)	2 (10%)	3 (15%)
61 - 70	1 (5%)	2 (10%)	3 (15%)
71 - 80	4 (20%)	1 (5%)	5 (25%)
81 - 90	3 (15%)	2 (10%)	5 (25%)
91 - 100	1 (5%)	3 (15%)	4 (20%)
Total	10 (50%)	10 (50%)	20 (100%)

Looking at this table gives me very useful information. I know that out of the 5 students with grades in the 80s, 3 are male and 2 are female, and out of the 4 students with grades in the 90s, 1 is male and 3 are female.

To get even more useful information, I can also include the percentages (relative frequencies), as in this table. I know that out of 25 percent of students with grades in the 70s, 20 percent are male and 5 percent are female. This useful information would have taken me a long time to find if I didn't organize the data into the contingency table.

A **contingency table** lists the actual and relative frequencies of two variables at the same time. Contingency tables are also known as cross-tabulations or cross-tabs.

Charting Your Course: Graphs

Charts are yet another efficient way to summarize and display patterns in a set of data. There are several forms of graphs: histograms, bar charts, pie charts, and line charts. So let's take a look at each one of them.

Histogram

A *histogram* is a graph to turn your frequency distribution table into something even more visual. It is a special type of bar chart that plots classes on the horizontal axis and frequencies on the vertical axis. The height of each bar represents the number of observations (frequencies) in each class. The histogram does not have gaps between its bars since the classes shown on the horizontal axis are continuous. Figure 3.1 shows the histogram for the students' grades in the previous example.

A **histogram** is a bar graph showing the number of observations in each class as the height of each bar.

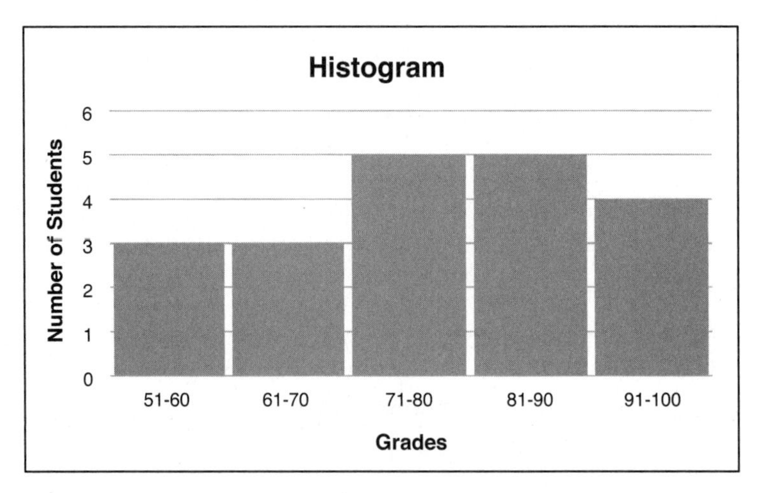

Figure 3.1
A histogram of students' grades.

Letting Excel Do Our Dirty Work

Excel can actually construct the frequency distribution for us and then plot the histogram. How nice!

Before we start, make sure you have the Data Analysis add-in installed on your computer. If you don't see this option in your Data tab, then see the section "Installing the Data Analysis Add-In" from Chapter 2. Now let's put Excel to work and create the frequency distribution and histogram for our grades example. Follow these steps:

1. Open a blank Excel sheet, and in cell A1 type the name of the variable (in our example, "Grades"). Starting in cell A2, enter all the raw data (in our example, each grade).
2. In cell B1, again type the name of the variable ("Grades"). Starting in cell B2, enter the upper limits of each class (see Figure 3.2). For example, in the class 51-60, the upper limit would be 60. Excel refers to the classes as bins.

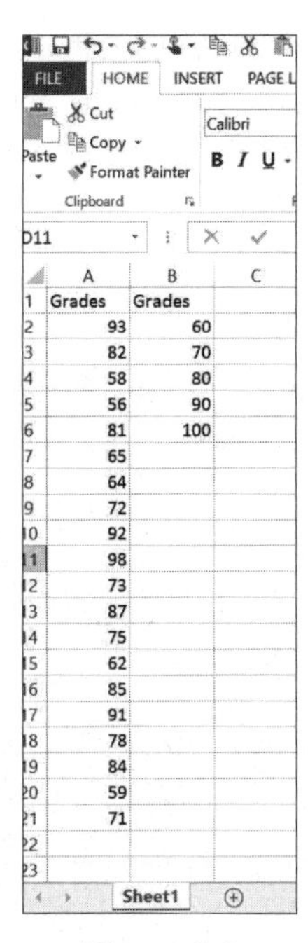

Figure 3.2

Raw data for the frequency distribution.

3. Go to the Data tab at the top of the Excel window, click on Data Analysis, and select Histogram (see Figure 3.3). Click the OK button.

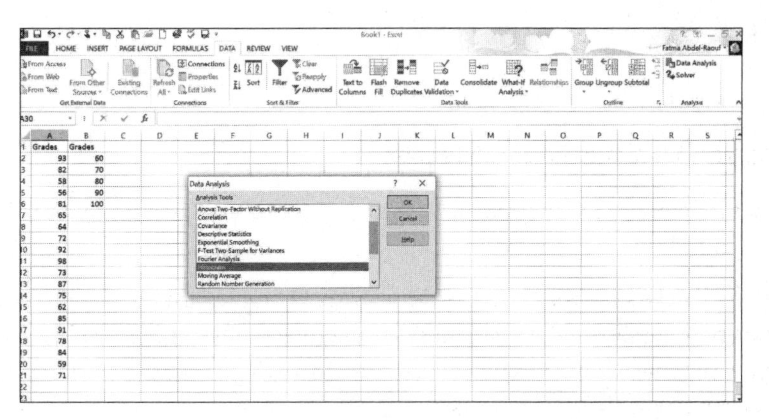

Figure 3.3

Excel Data Analysis dialog box.

4. The Histogram dialog box (see Figure 3.4) will appear. Click in the Input Range list box, and then click in the worksheet to select the data cells and the column label (cells A1

through A21 in our example). Then click in the Bin Range list box, and in the worksheet select the upper limit cells and the column label (cells B1 through B6 in our example).

Histogram
Input
Input Range: $A$1:$A$21
Bin Range: $B$1:$B$6
Labels
Output options
Output Range:
New Worksheet Ply:
New Workbook
Pareto (sorted histogram)
Cumulative Percentage
Chart Output
OK
Cancel
Help

Figure 3.4
Excel Histogram dialog box.

5. Click the Labels check box since we typed the name of the variable in the first cells, A1 and B1. (This way, Excel will display the labels you typed on the output graph instead of having to add them manually, later.) Click the Chart Output check box.

6. Click the OK button to generate the frequency distribution and the histogram (see Figure 3.5).

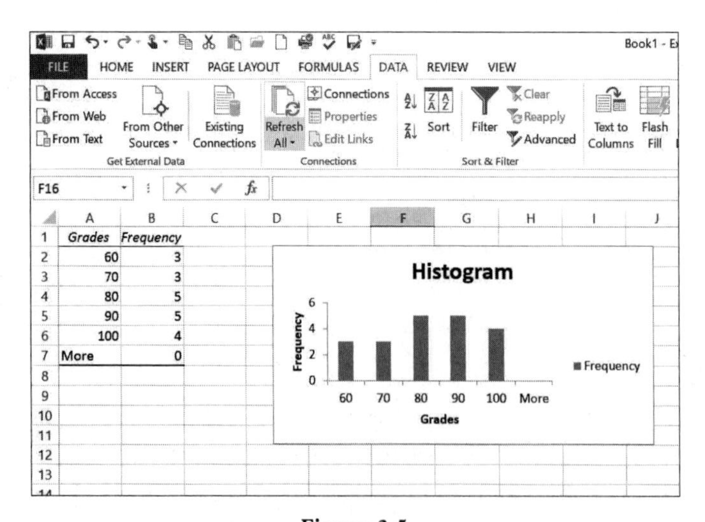

Figure 3.5
Frequency distribution and histogram.

Excel does a good job with the histogram, but we need to make two changes:

1. Go to the table and clear the cells with "more" and "0" in them. Look at what happens in the graph! Cool isn't it? Excel removed the word from the graph, also (see Figure 3.6).

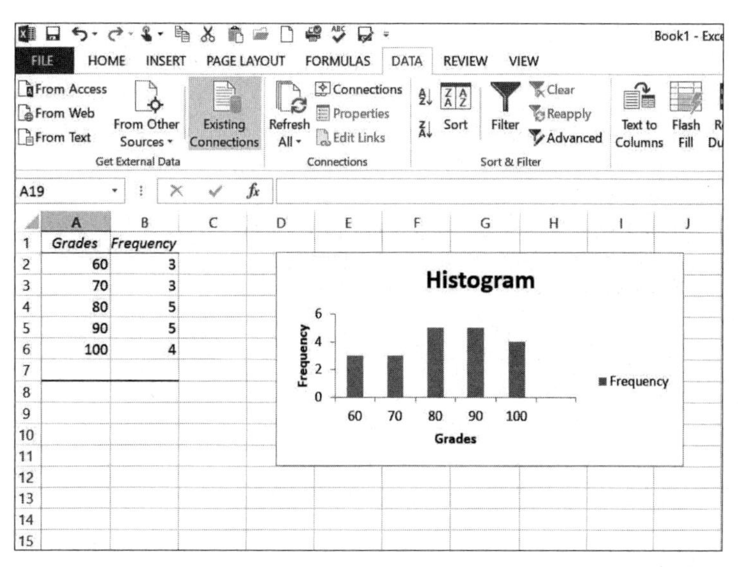

Figure 3.6

Removing "More" from the histogram.

2. Because the histogram doesn't have gaps between its bars, we should remove them. Highlight the bars in the graph, right click, and select Format Data Series.... A dialog box will appear, as shown in Figure 3.7.

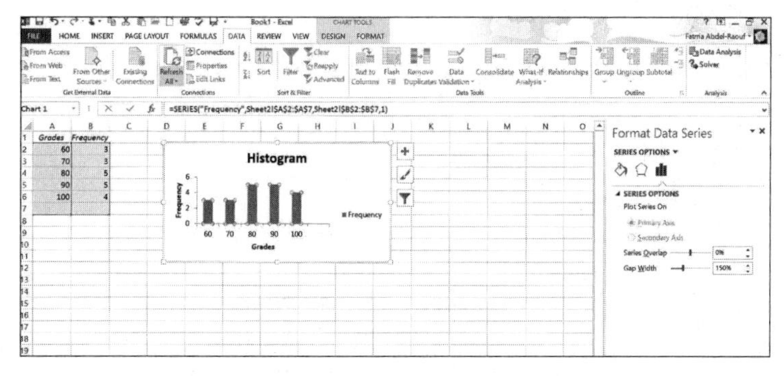

Figure 3.7

Removing the gaps between bars on the histogram.

3. Reduce the Gap Width percentage to a single digit to show that the classes are continuous. I reduce it to 4 percent in our example (see Figure 3.8).

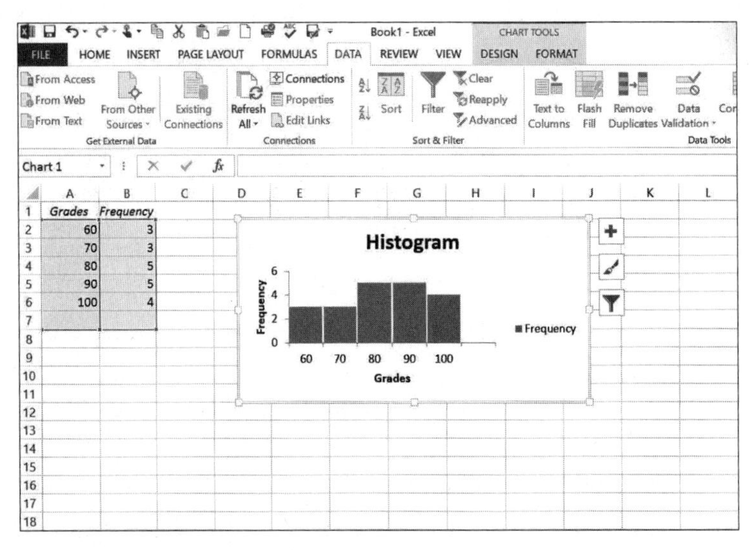

Figure 3.8

Final histogram.

Check it out! Now your frequency distribution and the histogram are final. Thanks Excel!

Bar Chart

A bar chart is useful when you're plotting individual data values next to each other. To demonstrate this type of chart (see Figure 3.9), we'll use the data from the following table, which represents the monthly credit card balances for an unnamed spouse of an unnamed person writing a statistics book. (Somebody is going to be in *big* trouble when she sees this.)

Anonymous Credit Card Balances

Month	Balance ($)
1	375
2	514
3	834
4	603
5	882
6	468

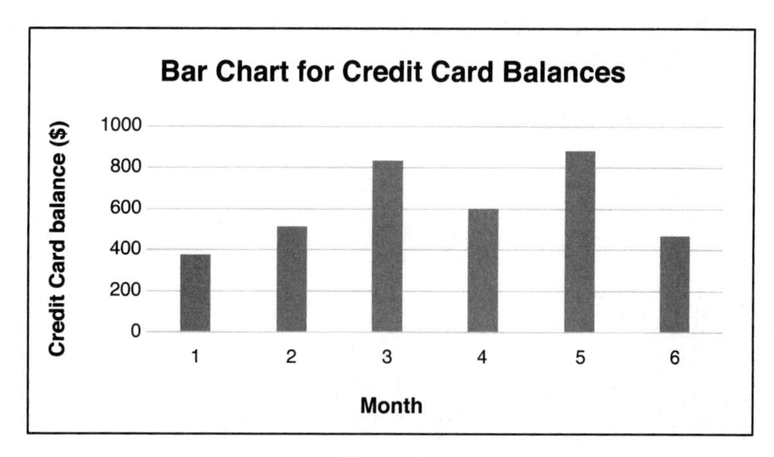

Figure 3.9
Bar chart for somebody's credit card balances.

You might be thinking now, "How can I get Excel to help me out?" Here is how:

1. Enter the labels and the data in columns A and B as shown in Figure 3.10.

	A	B	C	D
1	Month	Balance ($)		
2	1	375		
3	2	514		
4	3	834		
5	4	603		
6	5	882		
7	6	468		
8				
9				
10				
11				
12				
13				

Figure 3.10
Data entered in Excel.

2. Highlight the balance data and label, cells B1 to B7. Go to the Insert tab at the top of Excel window, select the Column chart, and from the menu choose the first graph under 2-D Column. Excel will output the graph shown in Figure 3.11.

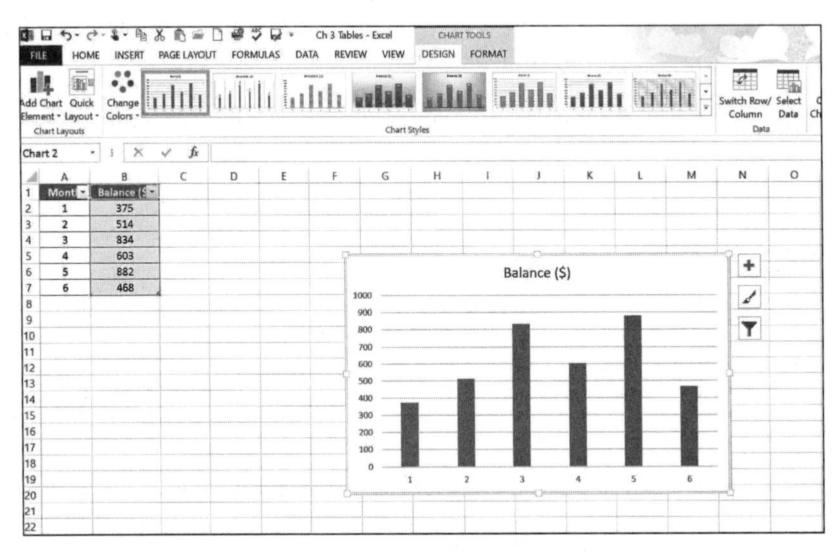

Figure 3.11
Excel bar chart.

3. Click on the plus sign next to the chart for the Chart Elements dialog, and select the Axis Titles checkbox to add labels to the axes (see Figure 3.12). On the horizontal axis, type "Month" and on the vertical axis, type "Credit Card Balance ($)." For the chart title, delete "Balance" and type "Bar Chart for Credit Card Balances."

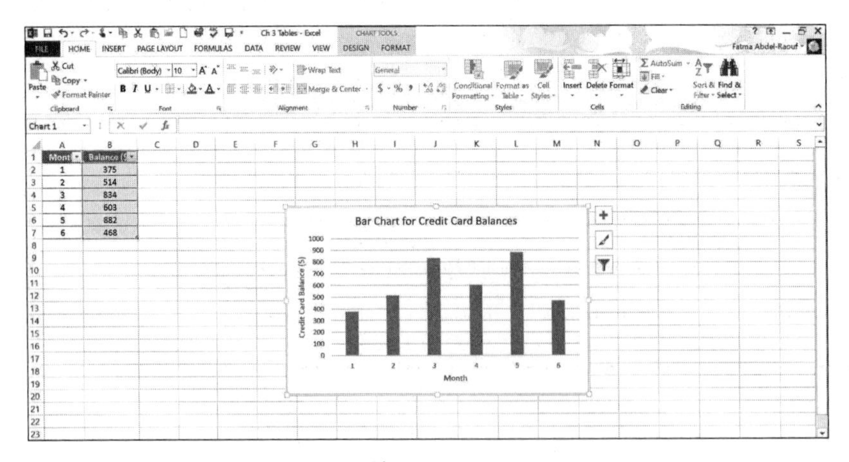

Figure 3.12
A fine-tuned bar chart in Excel.

Now you have a nice looking bar chart.

RANDOM THOUGHTS

Now you might be asking yourself "Why did I choose the Column chart and not the Bar chart in Excel when I'm trying to create a bar chart?" Excel simply uses different names; When the bars are vertical, Excel calls it a Column chart, and when the bars are horizontal, Excel calls it a Bar chart. Go figure!

TEST YOUR KNOWLEDGE

What is the difference between a histogram and a bar chart? There are two main differences:

1. For the histogram, you have to present the classes on the horizontal axis and the frequency on the vertical axis, whereas for the bar chart, you can present any variable on the axes.
2. The histogram has no gaps between the bars, whereas the bar chart does have gaps between the bars. Also, in the bar chart, bars can be represented vertically or horizontally.

Pie Chart

Pie charts are commonly used to describe data from relative frequency distributions. This type of chart is simply a circle divided into portions whose area is equal to the relative frequency distribution. Pie charts are used extensively in statistics, as they show the importance of a part (or a wedge of the circle) relative to the whole. Let's use an example to illustrate it. An anonymous statistics professor submitted the following final grade distribution:

Grade	Number of Students	Relative Frequency
A	9	$9/30 = 30\%$
B	13	$13/30 = 43\%$
C	6	$6/30 = 20\%$
D	2	$2/30 = 7\%$
	Total = 30	= 100%

We can present these data using a pie chart, shown in Figure 3.13.

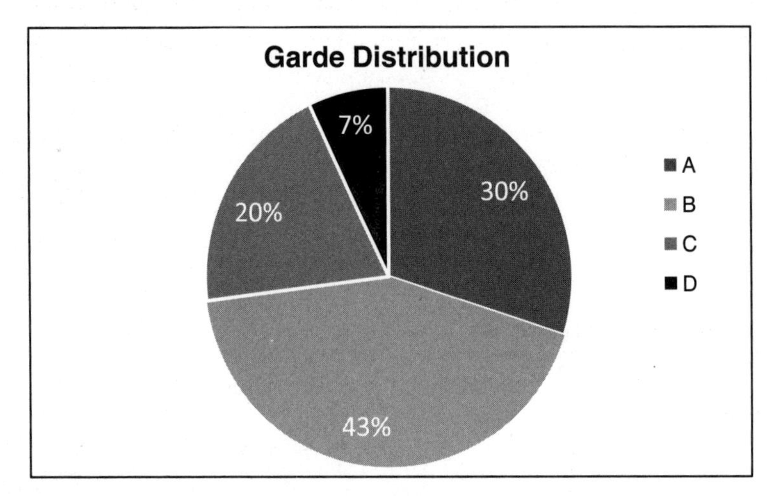

Figure 3.13
A pie chart illustrating a grade distribution.

I know you must be wondering if Excel can do this for you. Yes, Excel can easily help. Type the data and labels in columns A and B. Highlight the data in cells A1 to B5, select the Insert tab at the top of Excel window, and choose the Pie chart. You get this nice looking pie chart!

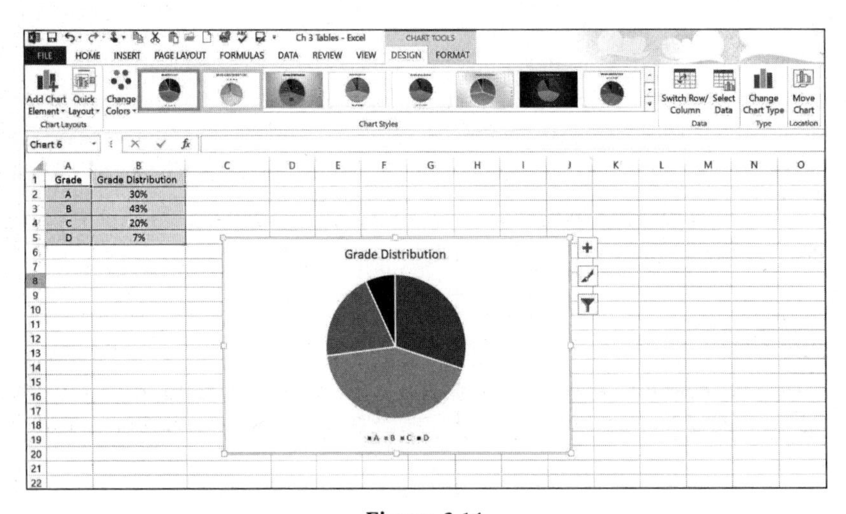

Figure 3.14
Pie chart in Excel.

We need to add a few things to make it look nicer and more informative. Click on the plus sign next to the chart, and select the Data Labels checkbox. You can also move around the legend; I like to have it on the right instead of at the bottom! Just double click on the Legend and choose a different placement. You will get this nice looking pie chart shown in Figure 3.15!

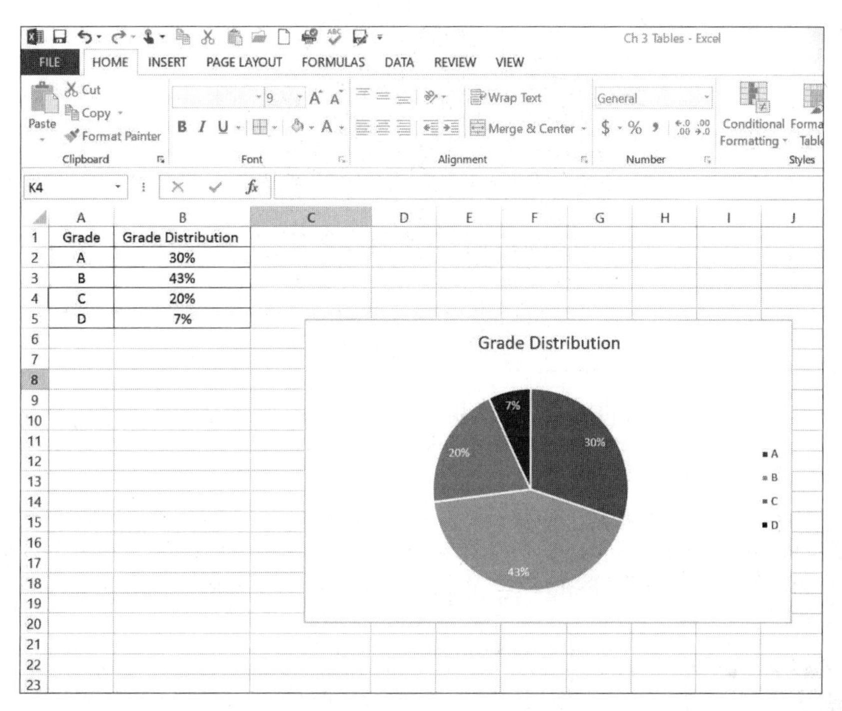

Figure 3.15

A final pie chart in Excel.

As you can see, the pie chart is much easier to interpret compared to the data in the table. This person must be a pretty good statistics teacher!

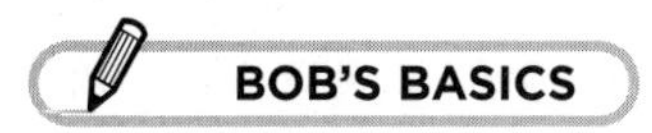

Pie charts are an excellent way to colorfully present data from a relative frequency distribution. Also use patterns and textures to distinguish the different slices.

To construct a pie chart by hand, you first need to calculate the *center angle* for each slice in the pie, which is illustrated in Figure 3.16.

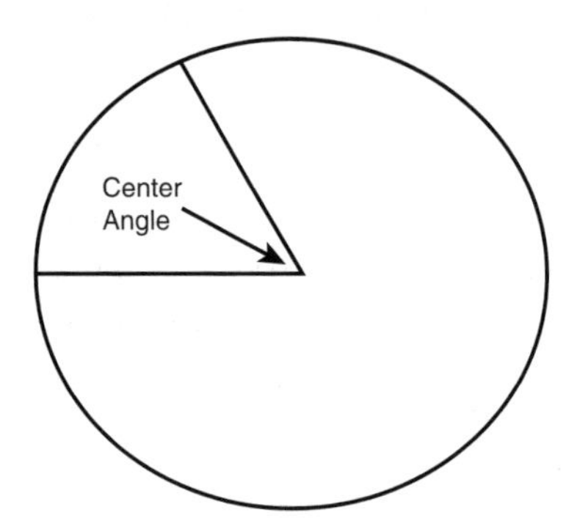

Figure 3.16
The center angle of a pie chart slice.

You determine the center angle of each slice by multiplying the relative frequency of the class by 360 (which is the number of degrees in a circle). These results are shown in the following table.

Center Angle for Pie Chart Construction

Grade	Relative Frequency	Central Angle
A	$^9/_{30} = 0.30$	$0.30 \bullet 360 = 108°$
B	$^{13}/_{30} = 0.43$	$0.43 \bullet 360 = 155°$
C	$^6/_{30} = 0.20$	$0.20 \bullet 360 = 72°$
D	$^2/_{30} = 0.07$	$0.07 \bullet 360 = 25°$
	Total = 1.00	= 360°

By using a device to measure angles, such as a protractor, you can now divide your pie chart into slices of the appropriate size. This assumes, of course, that you've mastered the art of drawing circles.

TEST YOUR KNOWLEDGE

Who was the first one to create the pie chart? William Playfair (a businessman, engineer, and economics writer from Scotland) created it in 1801 in his publication "The Statistical Breviary."

I'm sure your inquisitive mind is now screaming with the question "How do I choose between a pie chart and a bar chart?" If your objective is to compare the relative size of each class to one another, use a pie chart. Bar charts are more useful when you want to highlight the actual data values.

Line Chart

A line chart is used to help identify patterns between two sets of data. To illustrate the use of line charts, we will use the following unemployment data in the United States from 2005 to 2015.

Year	Unemployment Rate, January of each Year
2005	5.3
2006	4.7
2007	4.6
2008	5.0
2009	7.8
2010	9.8
2011	9.2
2012	8.3
2013	8.0
2014	6.6
2015	5.7

To be able to see the pattern in the unemployment rate in the United States from 2005 to 2015, we can plot the data on a line chart, which is shown in Figure 3.17.

The line chart shows an increase in the unemployment rate from 2008 to 2010 (this is a reflection of the Great Recession in the U.S. economy at that time), and then the rate decreases.

I see you thinking about how you can probably create this with Excel the same way you did the bar chart and the pie chart. Yes, you are right. Enter the data and select a line chart from the Insert tab. Excel is very helpful, isn't it?

RANDOM THOUGHTS

When you draw any graph, don't forget to label the axes. This is a must do! The same graph with different labels on the axes can represent an entirely different data relationship.

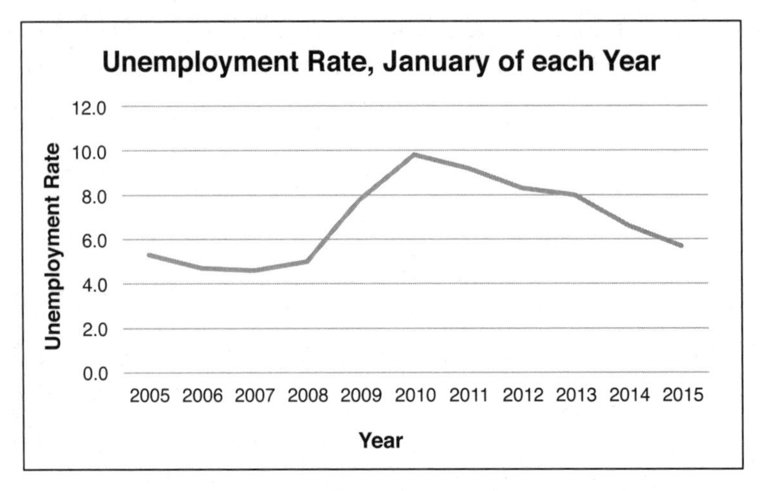

Figure 3.17
A line chart representing the unemployment rate in the United States.

The Stem and Leaf Display—Statistical Flower Power

The *stem and leaf display* is another graphical technique you can use to display your data. A statistician named John Tukey originated the idea during the 1970s. The major benefit of this approach is that all the original data points are visible on the display.

The **stem and leaf display** splits the data values into stems (the first digit or digits in the value) and leaves (the remaining digit or digits in the value). By listing all of the leaves to the right of each stem, we can graphically describe how the data is distributed.

To demonstrate this method, we will use Bob's son Brian's golf scores for his last 24 rounds, shown in the following table. Normally, Brian would only report his better scores, but we statisticians must be unbiased and accurate.

Brian's Golf Scores							
81	86	78	80	81	82	92	90
79	83	84	95	85	88	80	78
84	79	80	83	79	87	84	80

Figure 3.18 shows the stem and leaf display for these scores.

7|88999
8|00001123344445678
9|025

Stem and Leaf Display

Figure 3.18
Stem and leaf display.

The "stem" in the display is the first column of numbers, which represents the first digit of the golf scores. The "leaf" in the display is the second digit of the golf scores, with 1 digit for each score. Because there were 5 scores in the 70s, there are 5 digits to the right of 7.

If we choose to, we can break this display down further by adding more stems. Figure 3.19 shows this approach.

7 (5)|88999
8 (0)|000011233444
8 (5)|5678
9 (0)|02
9 (5)|5

More Detailed Stem and Leaf Display

Figure 3.19
A more detailed stem and leaf display.

Here, the stem labeled 7 (5) stores all the scores between 75 and 79. The stem 8 (0) stores all the scores between 80 and 84. After examining this display, I can see a pattern that's not as obvious when looking at Figure 3.18: Brian usually scores in the low 80s.

Now that you have mastered the art of displaying descriptive statistics, you are ready to move on to calculating them in the next chapter.

Practice Problems

1. The following table represents the exam grades from 36 students from a certain class that I taught. Construct a frequency distribution with 9 classes ranging from 56 to 100.

Exam Scores

60	95	75	84	85	74
81	99	89	58	66	98
99	82	62	86	85	99
79	88	98	72	72	72
75	91	86	81	96	86
78	79	83	85	92	68

2. Construct a histogram using the solution from Problem 1.
3. Construct a relative and a cumulative frequency distribution from the data in Problem 1.
4. Construct a pie chart from the solution to Problem 1.
5. Construct a stem and leaf display from the data in Problem 1 using one stem for the scores in the 50s, 60s, 70s, 80s, and 90s.
6. Construct a stem and leaf display from the data in Problem 1 using two stems for the scores in the 50s, 60s, 70s, 80s, and 90s.

The Least You Need to Know

- Frequency distributions are an efficient way to summarize data by counting the number of observations in various groupings.
- Histograms provide a graphical overview of data from frequency distributions.
- Pie, bar, and line charts are effective ways to present data in different graphical forms.
- Stem and leaf displays not only provide a graphical display of the data's distribution, but they also contain the actual data values of interest.

CHAPTER

4

Calculating Descriptive Statistics: Measures of Central Tendency (Mean, Median, and Mode)

In This Chapter

- Understanding central tendency
- Calculating the mean, median, mode, and weighted mean of a sample and population
- Calculating the mean of a frequency distribution
- Using Excel to calculate central tendency

The emphasis of Chapter 3 was to demonstrate ways to display our data graphically so that our brain cells could quickly absorb the big picture. With that task behind us, we can now proceed to the next step—summarizing our data numerically. This chapter allows us to throw around some really cool words like "median" and "mode" and, when we're through, you'll actually know what they mean!

As mentioned in Chapter 1, descriptive statistics form the foundation for practically all statistical analysis. If these are not calculated with loving care, our final analysis could be misleading. And as everybody knows, statisticians never want to be misleading. So this chapter focuses on how to calculate descriptive statistics manually and, if you so choose, how to verify these results with our good friend Excel.

This is the first chapter that uses mathematical formulas that have all those funny-looking Greek symbols that can make you break out into a cold sweat. But have no fear. We will slay these demons one by one through careful explanation and, in the end, victory will be ours. Onward!

Measures of Central Tendency

There exist two broad categories of descriptive statistics that are commonly used. The first, *measures of central tendency*, describes the center point of our data set with a single value. It's a valuable tool to help us summarize many pieces of data with one number. The second category, *measures of dispersion*, is the topic of Chapter 5. But let's explore the many ways to measure the central tendency of our data.

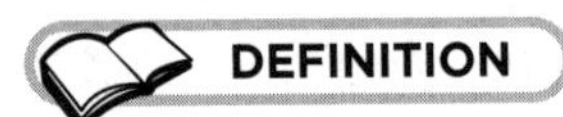

Measures of central tendency describe the center point of a data set with a single value. **Measures of dispersion** describe how far individual data values have strayed from the mean.

We will start by looking at the measures of central tendency for ungrouped data and then move to grouped data.

What is the difference between ungrouped data and grouped data? **Ungrouped data** is when you have the actual observations in raw data form, whereas **grouped data** is when you have a frequency distribution to present the data, as the ones illustrated in the previous chapter. As an example, the grades for 10 students in my statistics class are

60 80 70 65 70 85 63 75 68 82;

this is ungrouped data. If I present the same data to you in a frequency distribution, such as the ones we did in Chapter 3, then it is grouped data.

Measures of central tendency for ungrouped data include: mean, median, mode, and weighted mean. Let's look at each one of them and see how to calculate it.

Mean

The most common measure of central tendency is the *mean* or *average*, which we calculate by adding all the values in our data set and then dividing this result by the number of observations. The mathematical formula for the mean differs slightly depending on whether you're referring to the sample mean or the population mean. The formula for the sample mean is as follows:

$$\overline{x} = \frac{\sum_{i=1}^{n} x_i}{n}$$

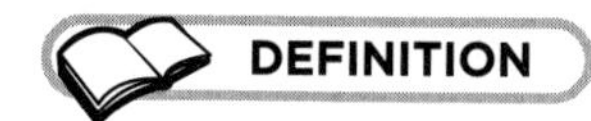

The **mean** or **average** is the most common measure of central tendency and is calculated by adding all the values in our data set and then dividing this result by the number of observations.

where:

$\bar{x}$ = the sample mean

x_i = the values in the sample (x_1 = the first data value, x_2 = the second data value, and so on)

$\sum_{i=1}^{n} x_i$ = the sum of all the data values in the sample

n = the number of data values in the sample

BOB'S BASICS

Don't panic when you see the symbol $\sum_{i=1}^{n} x_i$, which means "the sum of x_i for $i = 1$ to n." If our data sample contains the values 5, 8, and 2, then $n = 3$, $x_1 = 5$, $x_2 = 8$, and $x_3 = 2$, resulting in the expression:

$$\sum_{i=1}^{n} x_i = x_1 + x_2 + x_3 = 5 + 8 + 2 = 15$$

The formula for the population mean is as follows:

$$\mu = \frac{\sum_{i=1}^{N} x_i}{N}$$

where:

μ = the population mean (pronounced *mu*, as in "I hope you find this a*mu*sing")

$\sum_{i=1}^{N} x_i$ = the sum of all the data values in the population

N = the number of data values in the population

To demonstrate calculating measures of central tendency, let's use the following example. As in many teenagers' households, video games are a common form of entertainment in Bob's family room. The following data set represents the number of hours each week that video games are played in Bob's household.

3	7	4	9	5	4	7	17	4	7

Because this data represents a sample, we will calculate the sample mean:

$$\overline{x} = \frac{\sum_{i=1}^{n} x_i}{n} = \frac{3+7+4+9+5+4+6+17+4+7}{10} = 6.6 \text{ hours/week}$$

You are probably saying, "This is not too many hours. I play more!"

Median

Another way to measure central tendency is by finding the *median*. The median is the observation in the middle of the distribution for which half of the observations are higher and half of the observations are lower. We find the median by arranging the data in ascending order and identifying the halfway point. The following median position formula helps you to identify the observation in the middle of your distribution.

Median position = $\frac{n+1}{2}$, where n is the number of observations.

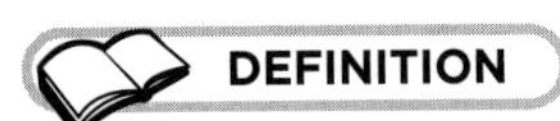

DEFINITION

The **median** is a measure of central tendency that represents the observation in the middle of the data set for which half of the observations are higher and half of the observations are lower.

Using a slightly modified data set for video game hours per week, we rearrange our observations in ascending order:

3 4 4 4 5 7 7 7 9 17

Apply the median position formula: median position = $\frac{10+1}{2}$ = 5.5.

Because we have an even number of data points (10 observations), the median position is the average of the 5th and 6th observations, which are 5 and 7. So, the median is 6 hours of video games per week. Notice that there are five data values to the left of the median (3, 4, 4, 4, and 5) and five data values to the right (7, 7, 7, 9, and 17). As you can see, 6 is the observation exactly in the middle of the distribution.

To illustrate the median for a data set with an odd number of values, let's remove 17 from the video games data and repeat our analysis.

3 4 4 4 5 7 7 7 9

Now, the median position = $\frac{9+1}{2}$ = 5. So the median is the 5th observation, which is also 5 hours of video games per week. (This is only a coincidence that the numbers happen to match!) Again, there are four data values to the left and right of this center point.

TEST YOUR KNOWLEDGE

The U.S. Census Bureau and other institutions report both the mean and the median in their data. According to the Census Bureau, in 2013 the median income for an individual with a Bachelor's degree is $50,738, while the mean is $62,048. The median income for a person with a high school degree is $30,286 and the mean is $35,309. So getting a college degree is worth the effort!

Mode

The mode is the observation in the data set that occurs the most frequently. A data set can have one mode, several modes, or even no mode.

To illustrate, let's use the following example of grades for 10 students in my statistics class:

60 80 70 65 70 85 60 75 60 82

The mode is 60 because this grade occurs three times more than any other score in our data set.

In this example, there is only one mode. Let's see another example where there is more than one mode. Grades for 10 students in my economics class are:

90 60 80 55 75 60 65 55 82 51

Here there are two modes: 55 and 60 since both of these values happen twice while all the rest happen only once. This is called "bimodal."

In other cases there can be no mode if all the observations happen with the same frequency. Let's illustrate it with the following examples of grades for 10 students in my finance class:

90 80 75 72 65 82 92 85 70 78

Here there is no mode since all of the observations happen once (Hint: Excel says "N/A" when there is no mode in the data).

Weighted Mean

When we calculated the mean number of hours of video games played in Bob's household, we gave each data value the same weight in the calculation as the other values. A *weighted mean* refers to a mean that needs to go on a diet. Just kidding; we were checking to see whether you were paying attention. A weighted mean allows you to assign more weight to certain values and less weight to others. For example, let's say your statistics grade this semester will be based on a combination of your exam grades and your homework grades, each weighted according to the following table:

Type	**Score**	**Weight (%)**
Midterm Exam	89	35%
Final Exam	95	50%
Homework	85	15%

We can calculate your final grade using the following formula for a weighted average. Note that here we are dividing by the sum of the weights rather than by the number of data values.

$$\overline{x} = \frac{\sum_{i=1}^{n} x_i w_i}{\sum_{i=1}^{n} w_i}$$

where:

w_i = the weight for each data value x_i

$\sum_{i=1}^{n} w_i$ = the sum of the weights

BOB'S BASICS

The symbol $\sum_{i=1}^{n} x_i w_i$ means "the sum of x times w." Each pair of x and w is first multiplied together, and these results are then summed.

The weighted mean equation can be set up in the following table to demonstrate the procedure.

Type (*i*)	**Score** (x_i)	**Weight** (w_i)	**Weight × Score** ($x_i w_i$)
Midterm Exam	89	0.35	31.15
Final Exam	95	0.50	47.50
Homework	85	0.15	12.75

$\sum_{i=1}^{3} w_i = 1.0$ $\sum_{i=1}^{3} x_i w_i = 91.4$

We can obtain the same result by plugging the numbers directly into the formula for a weighted average:

$$\overline{x} = \frac{(89)(0.35)+(95)(0.5)+(85)(0.15)}{0.35+0.5+0.15} = 91.4$$

Congratulations. You earned an A!

The weights in a weighted average do not need to add to 1 as in the previous example. Let's say I want a weighted average of my two most recent golf scores, 90 and 100, and I want 90 to have twice the weight as 100 in my average. I would calculate this by:

$$\overline{x} = \frac{(2)(90)+(1)(100)}{2+1} = 93.3$$

By giving more weight to my lower score, the result is lower than the true average of 95. In this case, I think I'll go with the weighted average.

Measures of Central Tendency for Grouped Data:

Here's some great news to get excited about! You can actually calculate the mean of grouped data from a frequency distribution. To illustrate, let's use our grade example from Chapter 3. We constructed the frequency distribution as follows:

Grades	Number of Students
51 - 60	3
61 - 70	3
71 - 80	5
81 - 90	5
91 - 100	4
	Total = 20

Remember, we call the intervals of the variable of interest the *classes*. To calculate the mean of grouped data, we first need to determine the midpoint of each class. Yes, there is a formula for that:

$$\text{Class midpoint} = \frac{\text{Lower Limit} + \text{Upper Limit}}{2}$$

For instance, the class midpoint for the first class in our example would be

$$\text{Class midpoint} = \frac{51+60}{2} = 55.5$$

The mean of the frequency distribution is given by the following equation—which is basically a weighted average formula:

$$\overline{x} = \frac{\sum_{i=1}^{k} f_i M_i}{\sum_{i=1}^{k} f_i}$$

where:

M_i = the midpoint for each class

f_i = the number of observations (frequency) of each class

k = the number of classes in the distribution

To calculate the mean in our example, we are going to add two columns: in the first column, we will calculate the midpoint of each class, and in the next column, we will multiply the frequency of each class by the midpoint. Add up the numbers in column *fM*, which gives us 1,550.

Grades	**Frequency (*f*)**	**Midpoint (*M*)**	***fM***
51 - 60	3	55.5	166.5
61 - 70	3	65.5	196.5
71 - 80	5	75.5	377.5
81 - 90	5	85.5	427.5
91 - 100	4	95.5	382
Total	20		1,550

Then applying the mean formula: $\overline{x} = \frac{1{,}550}{20} = 77.5$.

According to this frequency distribution, the average grade in the statistics class is 77.5. Not bad!

WRONG NUMBER

The mean of a frequency distribution where data is grouped into classes is only an approximation to the mean of the original data set from which it was derived. This is true because we make the assumption that the original data values are at the midpoint of each class, which is not necessarily the case. The true mean of the 20 original data values in our grade example is 75.8 rather than 77.5.

That wraps up all the different ways to measure central tendency in our data sets. However, one question is still screaming to be answered....

How Does One Choose?

I bet you never thought you would have so many choices of measuring central tendency! It's kind of like being in an ice cream store in front of 30 flavors. If you think all the data in your data set is relevant, then the mean is your best choice. This measurement is affected by both the number and magnitude of your values. However, very small or very large values can have a significant impact on the mean, especially if the size of the sample is small. If this is a concern, perhaps you should consider using the median. The median is not as sensitive to a very large or small value.

Consider the following video game hours per week data set:

3 4 4 4 5 6 7 7 9 17

The number 17 is rather large when compared to the rest of the data. The mean of this sample is 6.6, whereas the median is 5.5. If you think 17 is not a typical value that you would expect in this data set, the median would be your best choice for central tendency. The mode, on the other hand, can be used to describe data at the nominal scale level—that is, qualitative data.

Using Excel to Calculate Central Tendency

You might be thinking about whether Excel can calculate the measures of central tendency for you, and you are in luck! To illustrate how to use Excel for this, let's use the following example about the top ten richest Hollywood celebrities in 2015:

Name	Net Worth (in millions of dollars)
Tyler Perry	400
Bill Cosby	350
Adam Sandler	300
Leonardo DiCaprio	200
Julia Roberts	165
Drew Carey	165
Jennifer Aniston	150
Robert Downey Jr.	140
Drew Barrymore	125
Angelina Jolie	120

Source: itsgr9.com/top-10-richest-hollywood-celebrities

We only need the net worth amounts; I added the name column just in case you are curious. Two of my favorite celebrities are on here!

Here are the steps to calculate the measures of central tendency in Excel:

1. Open a blank Excel worksheet and enter the net worth data and label in column A (see Figure 4.1).

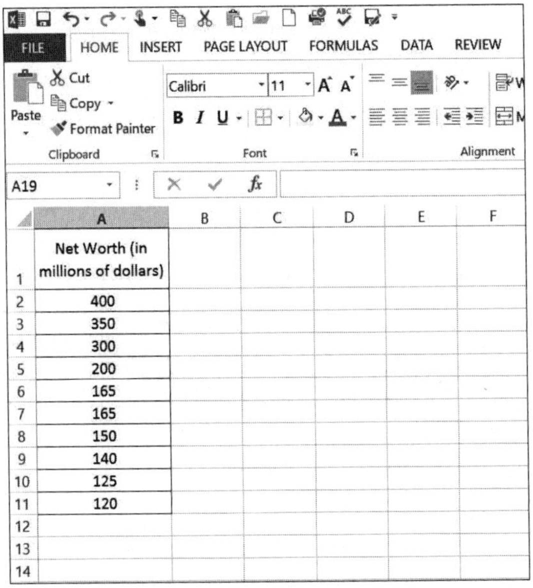

	A
1	Net Worth (in millions of dollars)
2	400
3	350
4	300
5	200
6	165
7	165
8	150
9	140
10	125
11	120

Figure 4.1

Enter the net worth data.

2. Click on the Data tab at the top of the Excel window and click on Data Analysis. (See the section "Installing the Data Analysis Add-In" in Chapter 2 for more details on this step if you don't see the Data Analysis option.) After selecting Data Analysis, you should see the dialog box shown in Figure 4.2.

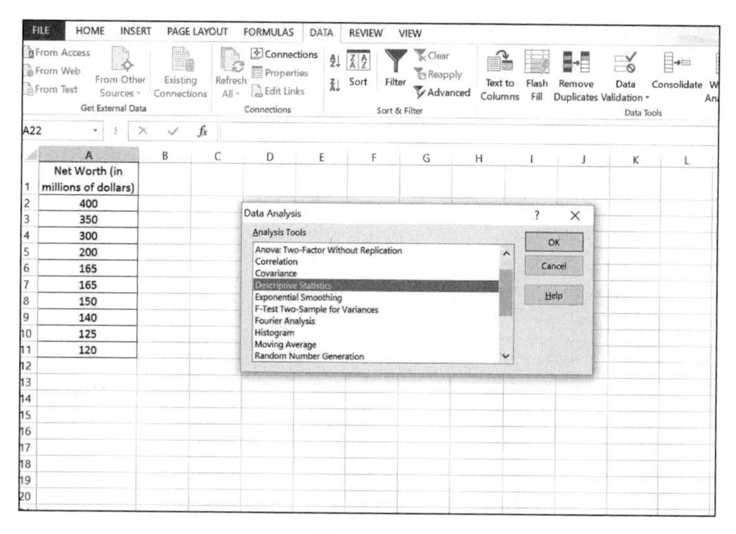

Figure 4.2
Data Analysis dialog box.

3. Select Descriptive Statistics and click the OK button. The following dialog box will appear (see Figure 4.3).

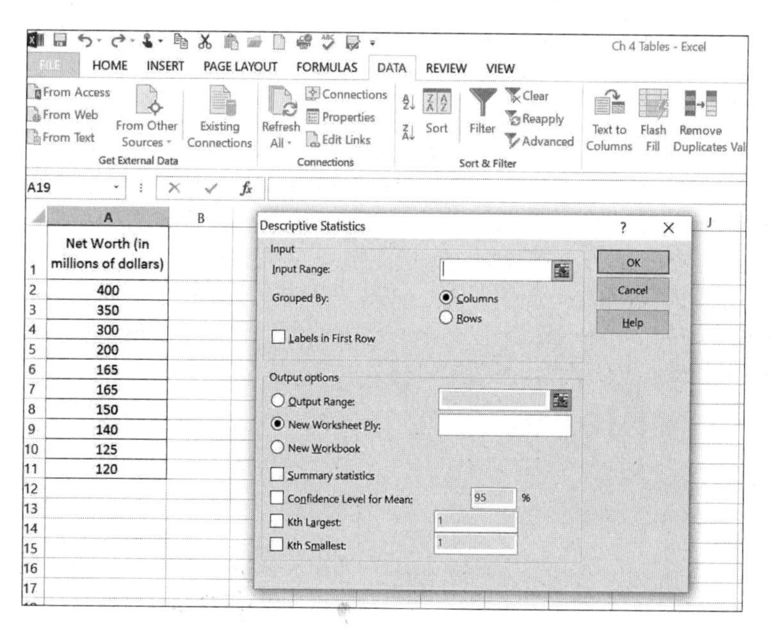

Figure 4.3
Descriptive Statistics dialog box.

4. For the Input Range, select cells A1 through A11. Check the Labels in First Row box since we have the name of the variable in the first cell. For output options, select Output Range and choose the cell where you want the output to be placed, C1 in our example. Then choose the Summary statistics check box and click OK.

5. After you expand columns C and D slightly to see all the figures, your spreadsheet should look like Figure 4.4.

	A	B	C	D
1	Net Worth (in millions of dollars)		*Net Worth (in millions of dollars)*	
2	400			
3	350		Mean	211.5
4	300		Standard Error	31.92917509
5	200		Median	165
6	165		Mode	165
7	165		Standard Deviation	100.9689171
8	150		Sample Variance	10194.72222
9	140		Kurtosis	-0.417122457
10	125		Skewness	1.056991807
11	120		Range	280
12			Minimum	120
13			Maximum	400
14			Sum	2115
15			Count	10

Figure 4.4

Measures of central tendency for the celebrity net worth example.

As you can see, the mean net worth is $211.5 million (Wow, right?), the median is $165 million, and the mode is $165 million. I wish that I were one of these celebrities!

Practice Problems

1. Calculate the mean, median, and mode for the following data set:
   20, 15, 24, 10, 8, 19, 24, 12, 21, 6

2. Calculate the mean, median, and mode for the following data set:
   84, 82, 90, 77, 75, 77, 82, 86, 82

3. Calculate the mean, median, and mode for the following data set:
   36, 27, 50, 42, 27, 36, 25, 40, 29, 15

4. Calculate the mean, median, and mode for the following data set:
   8, 11, 6, 2, 11, 6, 5, 6, 10

5. A company counted the number of their employees in each of the following age classes. According to this distribution, what is the average age of the employees in the company?

Age Range	Number of Employees
20–24	8
25–29	37
30–34	25
35–39	48
40–44	27
45–49	10

6. Calculate the weighted mean of the following values with the corresponding weights.

Value	Weight
118	3
125	2
107	1

The Least You Need to Know

- Calculate the mean of a data set by summing all the values and dividing this result by the number of values.
- A weighted mean allows you to assign more weight to certain values and less weight to others.
- The median of a data set is the midpoint of the set if the values are arranged in ascending or descending order.
- The median is the single center value from the data set if there is an odd number of values in the set. The median is the average of the two center values if the number of values in the set is even.
- The mode of a data set is the value that appears most often in the set. There can be more than one mode in a data set.

CHAPTER

5

Calculating Descriptive Statistics: Measures of Dispersion

In This Chapter

- Calculating the range of a sample
- Calculating the variance and standard deviation of a sample and population
- Calculating the variance and standard deviation of grouped data
- Using measures of relative position to identify outliers
- Using Excel to calculate measures of dispersion

In Chapter 4, we calculated measures of central tendency by summarizing our data set into a single value. But in doing so, we lost information that could be useful. For the video game example, if the only information I provided you was that the mean of my sample was 6.6 hours, you would not know whether all the values were between 6 and 7 hours or whether the values varied between 1 and 12 hours. As you will see later, this distinction can be very important.

To address this issue, we rely on the second major category of descriptive statistics, measures of dispersion, which describes how far the individual data values have strayed from the mean. So let's look closely at them and at the different ways we can measure dispersion.

Measures of dispersion tell me how the data are spread out around the mean. The smaller the value for the measure of dispersion is, the closer the data values are to the mean, whereas the larger the value for the measure of dispersion is, the more spread out the data values are.

To see it clearly, let's look at these two data sets:

1	3		5		7	9
	3	4	5	6	7	

Each data set has 5 observations and each has the same mean, which is 5. But the values in the first data set are more spread out, while the values in the second data set are closer to the mean. That's because the standard deviation (which is a measure of dispersion) is larger for the first data set than for the second.

In this chapter, we will look at measures of dispersion for ungrouped data, measures of dispersion for grouped data, and measures of relative position. So, let's get started.

Measures of Dispersion for Ungrouped Data

This includes range, variance, and standard deviation. Let's see what each one means.

Range

The *range* is the simplest measure of dispersion and is calculated by finding the difference between the highest value and the lowest value in the data set. To demonstrate how to calculate the range, I'll use the following example.

Obtain the **range** of a sample by subtracting the smallest value from the largest value.

One of Debbie's special qualities is that she is a dedicated grill-a-holic when it comes to barbequing in the backyard. The following data set represents the number of meals each month that Debbie cranks out on the turbo-charged grill:

7 9 8 11 4

The range of this sample would be:

$$\text{Range} = 11 - 4 = 7 \text{ meals}$$

As you can see, range is a very simple measure of dispersion. Yet it has a limitation—it relies on only two data values to describe the variation in the sample. No other values between the highest and lowest points are part of the range calculation. Variance and standard deviation, on the other hand, don't suffer from this limitation.

Variance

One of the most common measures of dispersion in statistics is the *variance*, which summarizes the squared deviation of each data value from the mean. The formula for the sample variance is:

$$s^2 = \frac{\sum_{i=1}^{n}(x_i - \bar{x})^2}{n-1}$$

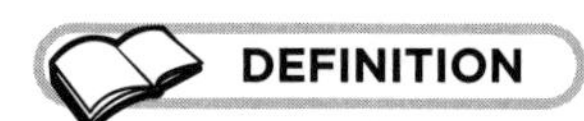

The **variance** is a measure of dispersion that describes the relative distance between the data points in the set and the mean of the data set. This measure is widely used in inferential statistics.

where:

s^2 = the variance of the sample

$\bar{x}$ = the sample mean

n = the size of the sample

$(x_i - \bar{x})$ = the deviation of each data value from the sample mean

The first step in calculating the sample variance is to determine the mean of the data set, which in the grilling example is 7.8 meals per month. The rest of the calculations can be facilitated by the following table.

x_i	$\bar{x}$	$(x_i - \bar{x})$	$(x_i - \bar{x})^2$
7	7.8	-0.8	0.64
9	7.8	1.2	1.44
8	7.8	0.2	0.04
11	7.8	3.2	10.24
4	7.8	-3.8	14.44

$$\sum_{i=1}^{5}(x_i - \bar{x})^2 = 26.80$$

The final sample variance calculation becomes this:

$$s^2 = \frac{26.80}{5-1} = 6.7$$

For those of us who like to do things in one step, we can also do the entire variance calculation in the following equation:

$$s^2 = \frac{(7-7.8)^2+(9-7.8)^2+(8-7.8)^2+(11-7.8)^2+(4-7.8)^2}{5-1} = 6.7$$

Using the Raw Score Method (When Grilling)

A more efficient way to calculate the variance of a data set is known as the *raw score method.* Even though at first glance this equation may look more imposing, its bark is much worse than its bite. Check it out and decide for yourself what works best for you.

$$s^2 = \frac{\sum_{i=1}^{n} x_i^2 - \frac{\left(\sum_{i=1}^{n} x_i\right)^2}{n}}{n-1}$$

where:

$\sum_{i=1}^{n} x_i^2$ = the sum of each data value after it has been squared

$\left(\sum_{i=1}^{n} x_i\right)^2$ = the square of the sum of all the data values

Okay, don't have heart failure just yet. Let me lay this out in the following table to prove to you there are fewer calculations than with the previous method.

x_i	x_i^2
7	49
9	81
8	64
11	121
4	16

$$\sum_{i=1}^{n} x_i = 39 \qquad \sum_{i=1}^{n} x_i^2 = 331$$

$$\left(\sum_{i=1}^{n} x_i\right)^2 = (39)^2 = 1{,}521$$

$$s^2 = \frac{331 - \frac{1521}{5}}{4} = \frac{331-304.2}{4} = 6.7$$

As you can see, the results are the same regardless of the method used. The benefits of the raw score method become more obvious as the size of the sample (n) gets larger.

BOB'S BASICS

If you are calculating the variance by hand, my advice is to do your fingers and calculator battery a favor and use the raw score method.

Variance of a Population

So far, we have discussed the variance in the context of samples. The good news is the variance of a population is calculated in the same manner as the sample variance. The bad news is we need to introduce another funny-looking Greek symbol: lowercase sigma. The equation for the variance of a population is as follows:

$$\sigma^2 = \frac{\sum_{i=1}^{N}(x_i-\mu)^2}{N}$$

where:

σ^2 = the variance of the population (pronounced "sigma squared")

$(x_i - \mu)$ = the deviation of each data value from the population mean

N = the size of the population

WRONG NUMBER

Notice that the denominator for the population variance equation is N, whereas the denominator for the sample variance is $n - 1$.

The raw score version of this equation is:

$$\sigma^2 = \frac{\sum_{i=1}^{N} x_i^2 - \frac{\left(\sum_{i=1}^{N} x_i\right)^2}{N}}{N}$$

Even though this procedure is identical to the sample variance, let me demonstrate with another example. Let's say I am considering my statistics class as my population and the following ages are the measurement of interest. (Can you guess which one is me? My age adds a little spice to the variance.)

21 23 28 47 20 19 25 23

I'll use the raw score method for this calculation with the population size (N) equal to 8. (I'd love to see a class this size.)

x_i	x_i^2
21	441
23	529
28	784
47	2,209
20	400
19	361
25	625
23	529

$$\sum_{i=1}^{N} x_i = 206 \qquad \sum_{i=1}^{N} x_i^2 = 5{,}878$$

$$\left(\sum_{i=1}^{N} x_i\right)^2 = (206)^2 = 42{,}436$$

$$\sigma^2 = \frac{5878 - \frac{42{,}436}{8}}{8} = \frac{5878 - 5{,}304.5}{8} = 71.7$$

Standard Deviation

This method is pretty straightforward. The *standard deviation* is simply the square root of the variance. Just as with the variance, there is a standard deviation for both the sample and population, as shown in the following equations.

A **standard deviation** is the square root of a variance.

Sample standard deviation:

$$s = \sqrt{s^2} = \sqrt{\frac{\sum_{i=1}^{n} (x_i - \bar{x})^2}{n-1}}$$

Population standard deviation:

$$\sigma = \sqrt{\sigma^2} = \sqrt{\frac{\sum_{i=1}^{N}(x_i - \mu)^2}{N}}$$

To calculate the standard deviation, you must first calculate the variance and then take the square root of the result. Recall from the previous sections that the variance from the sample of the number of meals Debbie grilled per month was 6.7. The standard deviation of this sample is as follows:

$$s = \sqrt{s^2} = \sqrt{6.7} = 2.6 \text{ meals}$$

Also recall the variance for the age of my class was 71.7. The standard deviation of the age of this population is as follows:

$$\sigma = \sqrt{\sigma^2} = \sqrt{71.7} = 8.5$$

The standard deviation is actually a more useful measure than the variance because the standard deviation is in the units of the original data set. In comparison, the units of the variance for the grill example would be 6.7 "meals squared," and the units of the variance for the age example would be 71.7 "years squared." I don't know about you, but I'm not too thrilled having my age reported as 2,209 squared years. I'll take the standard deviation over the variance any day.

Measures of Dispersion for Grouped Data

We start by calculating the variance of grouped data. (Remember, *grouped data* is when your data is presented as a frequency distribution instead of raw, ungrouped data.) Then to find the standard deviation, just take the square root of the variance.

There are different versions of the formula to calculate the variance for grouped data. This is the one I prefer:

$$S^2 = \frac{\sum_{i=1}^{k} fM^2 - n\overline{x}^2}{n-1}$$

Where:

f = the frequency in each class

M = the midpoint of each class

k = the number of classes

$\overline{x}$ = the sample mean

n = the number of observations

I like this formula because I think it's the simplest—just calculate the fM^2 column, as shown below, by multiplying the frequency by the square of the midpoint for each class. Add them up and you get the result of this: $\sum_{i=1}^{k} fM^2$. To illustrate, let's apply the formula to our grade example from the last chapter.

Grades	**Number of Students (*f*)**	**Midpoint (*M*)**	fM^2
51 - 60	3	55.5	$3(55.5)^2 = 9{,}240.75$
61 - 70	3	65.5	$3(65.5)^2 = 12{,}870.75$
71 - 80	5	75.5	$5(75.5)^2 = 28{,}501.25$
81 - 90	5	85.5	$5(85.5)^2 = 36{,}481$
91 - 100	4	95.5	$4(95.5)^2 = 36{,}481$
	Total = 20		= 123,645

In this example, $k = 5$ classes and $n = 20$ observations. We calculated the mean in Chapter 4, and it was 77.5 points. We just calculated $\sum_{i=1}^{5} fM^2$ in the table above and got 123,645. Substitute in the formula to get the variance as follows:

$$s^2 = \frac{123{,}645 - 20(77.5)^2}{20-1} = \frac{3{,}520}{19} = 185.26$$

Then take the square root of the variance to find the standard deviation, shown below.

The standard deviation, $s = \sqrt{s^2} = \sqrt{185.26} = 13.61$ points.

Measures of Relative Position

Another way of looking at dispersion of data is through measures of relative position, which describe the percentage of the data below a certain point. This technique includes quartile and interquartile measurements.

Quartiles

Quartiles divide the data set into four equal segments after it has been arranged in ascending order. Approximately 25 percent of the data points will fall below the *first quartile*, *Q*1. Approximately 50 percent of the data points will fall below the *second quartile*, *Q*2. And, you guessed it, 75 percent should fall below the *third quartile*, *Q*3. To demonstrate how to calculate *Q*1, *Q*2, and *Q*3, let's look at the following data showing the estimated population of the 10 largest cities in the United States from the 2010 Census.

Quartiles measure the relative position of the data values by dividing the data set into four equal segments.

City	Population (in millions)
New York, NY	8.18
Los Angeles, CA	3.79
Chicago, IL	2.7
Houston, TX	2.1
Philadelphia, PA	1.53
Phoenix, AZ	1.45
San Antonio, TX	1.33
San Diego, CA	1.31
Dallas, TX	1.2
San Jose, CA	0.95

Source: U.S. Census Bureau census.gov/2010census/popmap

1. Arrange the population data in an ascending order:

   0.95 1.20 1.31 1.33 1.45 1.53 2.10 2.70 3.79 8.18

2. Find the median of the data set. This is *Q*2.

   Median position = $\frac{10+1}{2} = 5.5$

   So $Q2 = \frac{1.45+1.53}{2} = 1.49$

3. Find the median of the lower half of the data (in parenthesis). This is *Q*1.

   (0.95 1.20 1.31 1.33 1.45) 1.53 2.10 2.70 3.79 8.18

   $Q1 = 1.31$

4. Find the median of the upper half of the data (in parenthesis), which is *Q*3.

   0.95 1.20 1.31 1.33 1.45 (1.53 2.10 2.70 3.79 8.18)

   $Q3 = 2.70$

Now we have all quartiles.

$Q1 = 1.31$

$Q2 = 1.49$

$Q3 = 2.70$

RANDOM THOUGHTS

California and Texas have three of their cities on the top 10 most populated cities in the United States according to the 2010 Census!

Interquartile Range (IQR)

When you have established the quartiles, you can easily calculate the *interquartile range* (IQR); the IQR measures the spread of the center half of our data set. It is simply the difference between the third and first quartiles, as follows:

$\text{IQR} = Q3 - Q1$

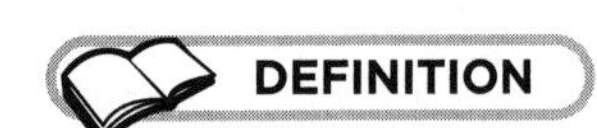

DEFINITION

The **interquartile range** measures the spread of the center half of the data set and identifies potential outliers. Outliers are observations with extreme values in the upper or lower quartiles and should be examined before you use them in analysis.

In our previous example of the top 10 most populated cities in the United States, IQR = 2.70 – 1.31 = 1.39. This is the range within which 50 percent of our data falls.

The interquartile range is used to identify outliers, which are the "black sheep" of our data sets. Outliers are observations with extreme values on either the upper or lower end of the data. Examine these outliers to determine how and why they appeared and whether similar values may continue to appear. You only need to discard outliers if they are mistakes and don't belong to the data set. For example, if you are looking at a GPA data set and you see an observation of 6.0, then it's definitely a mistake and does not belong (since the maximum GPA is 4.0). However, if the outliers belong to the data set, keep them even if they have extreme values, and they may help you learn something valuable about the data under investigation. John Tukey identified outliers as any value outside this range:

$Q1 - 1.5\ (\text{IQR})$ and $Q3 + 1.5\ (\text{IQR})$

So any value less than $Q1$ - 1.5 (IQR) or any value greater than $Q3$ + 1.5 (IQR) is considered an outlier. Let's apply this to our example of the top 10 most populated cities in the United States.

$$Q1 - 1.5\ (\text{IQR}) = 1.31 - 1.5\ (1.39) = -0.775$$

$$Q3 + 1.5\ (\text{IQR}) = 2.70 + 1.5\ (1.39) = 4.785$$

Looking at our data, we see that only one city of the population of the 10 most populated cities is outside of that range. New York City with its 8.18 million people is considered an outlier. However, it does belong to the data set—it's just an observation with an extremely large value compared to the rest of the data. If you have ever been to New York City, this won't surprise you!

TEST YOUR KNOWLEDGE

The IQR method for checking outliers was introduced by John Tukey (1915-2000), a famous American statistician. When asked why he uses the factor 1.5 in the outlier formula, Tukey answered, "because 1 is too small and 2 is too large."

Excel to the Rescue

I know that calculating the variance and standard deviation using the formula can be tedious and time consuming. Here comes Excel to help. Excel can calculate the range, variance, and standard deviation for you, and you don't need to complete any additional steps! Not even one! If we look at the Chapter 4 example from the section "Using Excel to Calculate Central Tendency" and look at the descriptive statistics table Excel provided us (shown again below, Figure 5.1), you will see that the table includes the range, variance, and standard deviation. So repeating those same steps, you can create a descriptive statistics table for any data set to analyze measures of dispersion.

As we can see from the previous figure, the range is $280 million, the sample variance is $10,194.72 and the standard deviation is $100.97 million. Piece of cake!

This wraps up our discussion on the different ways to describe measures of dispersion.

WRONG NUMBER

The values for variance and standard deviation reported by Excel are for a sample. If your data set represents a population, you need to recalculate manually the results using N in the denominator rather than $n - 1$.

	A	B	C	D
1	Net Worth (in millions of dollars)		*Net Worth (in millions of dollars)*	
2	400			
3	350		Mean	211.5
4	300		Standard Error	31.92917509
5	200		Median	165
6	165		Mode	165
7	165		Standard Deviation	100.9689171
8	150		Sample Variance	10194.72222
9	140		Kurtosis	-0.417122457
10	125		Skewness	1.056991807
11	120		Range	280
12			Minimum	120
13			Maximum	400
14			Sum	2115
15			Count	10

Figure 5.1

Measures of dispersion for the celebrity net worth example.

Practice Problems

1. Calculate the variance, standard deviation, and the range for the following sample data set:

   20, 15, 24, 10, 8, 19, 24

2. Calculate the variance, standard deviation, and the range for the following population data set:

   84, 82, 90, 77, 75, 77, 82, 86, 82

3. Calculate the variance, standard deviation, and the range for the following sample data set:

   36, 27, 50, 42, 27, 36, 25, 40

4. Calculate the quartiles and the cutoffs for the outliers for the following data set:

   8, 11, 6, 2, 11, 6, 5, 6, 10, 15

5. A company counted the number of their employees in each of the age classes as follows. According to this distribution, what is the standard deviation for the age of the employees in the company?

Age Range	Number of Employees
20–24	8
25–29	37
30–34	25
35–39	48
40–44	27
45–49	10

The Least You Need to Know

- The range of a data set is the difference between the largest value and smallest value.
- The variance of a data set summarizes the squared deviation of each data value from the mean.
- The standard deviation of a data set is the square root of the variance and is expressed in the same units as the original data values.
- The interquartile range measures the spread of the center half of the data set and identifies outliers, which are extreme values that need to be examined before using them in your analysis.

PART

2

Probability Topics

The connection between descriptive and inferential statistics is based on probability concepts. We know the topic of probability theory scares the living daylights out of many students, but it is a very important topic in the world of statistics. The topic of probability acts as a critical link between descriptive and inferential statistics. Without a firm grasp of probability concepts, inferential statistics will seem like a foreign language. Because of this, Part 2 is designed to help you over this hurdle.

CHAPTER

6

Introduction to Probability

As we leave the happy world of descriptive statistics, you may feel like you're ready to take on the challenge of inferential statistics. But before we enter that realm, we need to arm ourselves with probability theory. Accurately predicting the probability that an event will occur has widespread applications. For instance, the gaming industry uses probability theory to set odds for lotteries, card games, and sporting events.

The focus of this chapter is to start with the basics of probability, after which we will gently proceed to more complex concepts in Chapters 7 and 8. We'll discuss different types of probabilities and how to calculate the probability of simple events. We'll rely on data from frequency distributions to examine the likelihood of an event. So pull up a chair and let's roll those dice!

In This Chapter

- Distinguish between classical, empirical, and subjective probability
- Use frequency distributions to calculate probability
- Understand the relationship between events
- Demonstrate the intersection and union of events using a Venn diagram

What is Probability?

Probability is a measure of likelihood that an event will happen in the future. It takes a value between 0 and 1, often expressed as a percentage. The closer the probability is to 1, the more likely the event will happen in the future, and the closer the probability is to 0, the less likely the event will happen. Probability cannot be negative and cannot be greater than 1. The weather forecast is an example of probability, like when you hear in the news that there is a 60 percent chance of rain tomorrow. The higher the number is, the more likely it is to rain tomorrow, while the lower the number is, the less likely it is to rain tomorrow.

If you are absolutely sure that an event will occur, then the probability of this event occurring is 1. For example, in rolling a single die, what is the probability of getting a number less than 7? It is 1 because I'm 100 percent certain that the number I'll get is less than 7. It could be 1, 2, 3, 4, 5, or 6, and all of these are less than 7. On the contrary, the probability of an impossible event is zero. For example, in rolling the single die, what is the probability of getting a 7? There is no chance the number I roll can be 7, so it is an impossible event and its probability of occurring is 0.

Before we go any further, we need to tackle some new statistics jargon. The following terms are widely used when talking about probability:

- **Experiment** The process of measuring or observing an activity for the purpose of collecting data. An example is rolling a single die.
- **Outcome** A particular result of an experiment. An example is getting a head when flipping a coin.
- **Sample space** All the possible outcomes of the experiment. In rolling a single die, the sample space is {1, 2, 3, 4, 5, and 6}. Statistics people like to put { } around the sample space values because they think it looks cool.
- **Event** One or more outcomes of an experiment, which are a subset of the sample space. For example, in rolling a single die, getting an even number is an event.

Now that we know what probability is, let's look at the three main methods of measuring it: classical, empirical, and subjective.

Classical Probability

Classical probability refers to a situation when we know the number of possible outcomes of the event of interest and the total number of possible outcomes. It is based on the assumption that each outcome of the experiment is equally likely to occur. The probability of an event can be calculated using the following equation:

$$P(A) = \frac{\text{Number of possible outcomes in which Event A occurs}}{\text{Total number of possible outcomes in the sample space}}$$

where:

P(A): the probability that Event A will occur.

For example, what is the probability of randomly selecting a Jack from a deck of cards? P(Jack)= $\frac{4}{52}$. That's because we have 4 jacks in the deck and 52 cards all together. Now, let me test your solitaire skills. What is the probability of randomly selecting a face from a deck of cards? I hear you saying P(Face) = $\frac{12}{52}$. You are correct! I guess you play solitaire a lot!

Classical probability requires that you know the number of outcomes that pertain to a particular event of interest. You also need to know the total number of possible outcomes in the sample space.

To use classical probability, you need to understand the underlying process so you can determine the number of outcomes associated with the event. You also need to be able to count the total number of possible outcomes in the sample space. As you will see next, this may not always be possible.

Empirical Probability

When we don't know enough about the underlying process to determine the number of outcomes associated with an event, we rely on *empirical probability.* This type of probability observes the number of occurrences of an event through an experiment and calculates the probability from a relative frequency distribution. Therefore:

$$P(A) = \frac{\text{Frequency in which Event A occurs}}{\text{Total number of observations}}$$

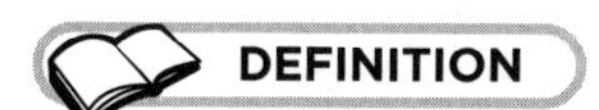

Empirical probability requires that you count the frequency that an event occurs through an experiment and calculate the probability from the relative frequency distribution.

One example of empirical probability is to answer the age-old question "What is the probability that John, Bob's son, will get out of bed in the morning for school after his first wake-up call?" Because Bob cannot understand the underlying process of why a teenager will resist getting out of bed before 2 P.M., he needs to rely on empirical probability. The following table indicates the number of wake-up calls John required over 20 school days.

John's Wake-Up Calls (Previous 20 School Days)

2	4	3	3	1	2	4	3	3	1
4	2	3	3	1	3	2	4	3	4

We can summarize this data with a relative frequency distribution.

Relative Frequency Distribution for John's Wake-Up Calls

Number of Wake-Up Calls	Number of Observations	Percentage
1	3	$^3/_{20} = 15\%$
2	4	$^4/_{20} = 20\%$
3	8	$^8/_{20} = 40\%$
4	5	$^5/_{20} = 25\%$
	Total = 20	

Based on these observations, if Event A = John getting out of bed on the first wake-up call, then P(A) = 0.15.

RANDOM THOUGHTS

The probability that you will win a typical state lottery, where you correctly choose 6 out of 49 numbers, is approximately 0.00000007, or 1 out of 14 million. This is calculated using classical probability. Compare this to the probability that you will be struck by lightning once during your lifetime (assume 80 years), which is 0.000083 or 1 out of 12,000 (source: www.lightningsafety.noaa.gov/odds.shtml).

If I choose to run another 20-day experiment of John's waking behavior, I would most likely see different results than those in the previous table. However, if I were to observe 100 days of this data, the relative frequencies would approach the true or classical probabilities of the underlying process. This pattern is known as the *law of large numbers.*

To demonstrate the law of large numbers, let's say I flip a coin three times and each time the result is heads. For this experiment, the empirical probability for the event heads is 100 percent. However, if I were to flip the coin 100 times, I would expect the empirical probability of this experiment to be much closer to the classical probability of 50 percent.

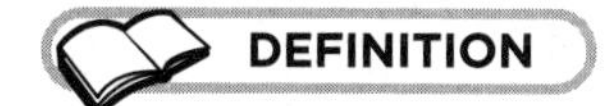

The **law of large numbers** states that when an experiment is conducted a large number of times, the empirical probabilities of the process will converge to the classical probabilities.

Subjective Probability

We use subjective probability when classical and empirical probabilities are not available. Under these circumstances, we rely on experience and intuition to estimate the probabilities.

Examples where we would apply subjective probability are "What is the probability that Bob's son Brian will ask to borrow Bob's new car, which happens to have a 6-speed manual transmission, for his junior prom?" (97 percent), or "What is the probability that Bob's new car will come back with all 6 gears in proper working order?" (18 percent). These probabilities are based on personal observations and experience. We need to use subjective probability in this situation because Bob's car would never survive several of these "experiments."

Relationship Between Events

To better understand probability, we need to know the meaning of three more statistics terms:

- Mutually Exclusive Events
- Independent Events
- Complementary Events

Mutually Exclusive Events

Events A and B are said to be *mutually exclusive* if they cannot occur at the same time during the experiment. For example, suppose my experiment is to roll a single die and my events are: Event A is to roll a 4 and Event B is to roll a 5. Can we get a 4 and a 5 simultaneously when we roll a single die once? No way, right? Since there is no way for these events to occur simultaneously, they are considered to be mutually exclusive. Another example is the result of your final exam as pass or fail. Can you pass and fail the same exam? No chance, so they are mutually exclusive.

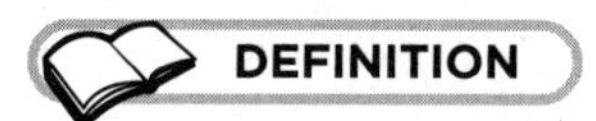

Two events are considered to be **mutually exclusive** if they cannot occur at the same time during the experiment.

Independent Events

Events A and B are said to be *independent* of each other if the occurrence of Event B has no effect on the probability of Event A. For example, suppose my experiment is to roll a single die twice and my events are: Event A is rolling a 2 and Event B is rolling a 6. The probability of getting a 6 when we roll the die for the second time is not affected by the probability of getting a 2 on the first roll. Therefore, these events are independent. Or for another example, in flipping a coin two separate times, the probability of getting a head on the second flip is not affected by the probability of getting a head on the first flip. Therefore, the events of getting a head the first time and a head the second time are independent events. On the other hand, dependent events depend on each other for their outcomes. For example, if your experiment is to draw marbles out of a bag and observe the color each time (without returning the marbles to the bag), then the probability of removing a certain color will change with each draw.

Events A and B are said to be **independent** of each other if the occurrence of Event B has no effect on the probability of Event A.

Complementary Events

The *complement* to Event A is defined as all the outcomes in the sample space that are not part of Event A. It is denoted as *A'* (pronounced A-prime), or *not A*. Since the probabilities of all events in the sample space add up to 1, then:

$P(A) + P(A') = 1$. Equivalently, we can also say $P(A') = 1 - P(A)$.

This rule is very useful in calculating probability. Let's apply it to some examples to see why. For instance, in rolling a single die, what is the probability of not getting a four? Not getting a 4 is the complement event of getting a 4, so $P(4') = 1 - P(4) = 1 - \frac{1}{6} = \frac{5}{6}$. Or in another example, in drawing a card from a deck, what is the probability of not getting a Jack? $P(\text{Jack}') = 1 - P(\text{Jack}) = 1 - \frac{4}{52} = \frac{48}{52}$. As you can see, using the complement rule makes the calculation easier and faster. Instead of having to count all the cards except Jacks, I can use the complement rule instead.

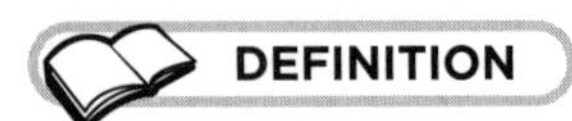

The **complement** to Event A is defined as all the outcomes in the sample space that are not part of Event A. It is denoted as A'.

Union and Intersection of Events

Understanding the union and intersection of events is important when using probability rules, as we will see in the next chapter. So before we delve into probability rules, let's see what these terms mean!

The Union of Events: A Marriage Made in Heaven

The *union* of Events A and B represents all the instances where *either* Event A *or* Event B *or both* occur and is denoted as $A \cup B$. For example, I'm teaching two undergraduate courses this semester: Statistics and Finance. Let's say that A represents students in my statistics class and B represents students in my finance class. $A \cup B$ includes all students who are in either my statistics class or in my finance class or in both. So any student in my statistics class but not in the finance class is part of the union. Likewise, any student in my finance class but not in the statistics class is part of the union. In addition, any student in both my statistics and my finance classes is part of the union.

The **union** of Events A and B represents all the instances where *either* Event A *or* Event B *or both* occur.

The Intersection of Events

The *intersection* of Events A and B represents all the instances where *both* Event A *and* Event B occur at the same time and is denoted as $A \cap B$. In our previous example, only students in both my statistics and finance classes are part of the intersection of the events. The probability of the intersection of two events is known as a *joint probability*.

The **intersection** of Events A and B represents all the instances where *both* Event A *and* Event B occur at the same time.

To better illustrate the union and intersection of events, let's look at the Venn diagram below.

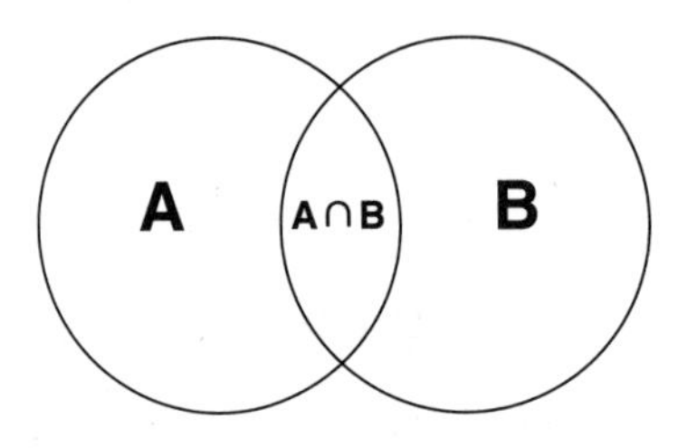

Figure 6.1
Venn diagram.

$A \cup B$ includes all elements in either A or in B or in both, so any element in the whole diagram is part of the union. $A \cap B$ includes all elements in both A and B, so only those elements at the intersection of the two circles are part of the intersection.

The Venn diagram was developed by English mathematician John Venn (1834–1923).

Practice Problems

1. Define each of the following as classical, empirical, or subjective probability.
   a. The probability that the baseball player Derek Jeter will get a hit during his next bat.
   b. The probability of drawing an Ace from a deck of cards.
   c. The probability that I will shoot lower than a 90 during my next round of golf.
   d. The probability of winning the next state lottery drawing.
   e. The probability that I will finish writing this book before my deadline.

2. Identify whether each of the following are valid probabilities.

    a. 65%

    b. 1.9

    c. 110%

    d. -4.2

    e. 0.75

    f. 0

The Least You Need to Know

- Classical probability requires knowledge of the underlying process in order to count the number of possible outcomes of the event of interest.
- Empirical probability relies on historical data from a frequency distribution to calculate the likelihood that an event will occur.
- The law of large numbers states that when an experiment is conducted a large number of times, the empirical probabilities of the process will converge to the classical probabilities.
- The union of Events A and B represents the number of instances where either Event A or B or both occur.
- The intersection of Events A and B represents the number of instances where Events A and B occur at the same time.

CHAPTER

7

Probability and Counting Rules

Now that we have arrived at the second of three basic probability chapters, we're ready for some new challenges. We need to take the probability concepts that you've mastered from Chapter 6 and put them to work on the next step up the ladder. Don't worry if you're afraid of heights like I am—just keep looking up!

This chapter deals with the topic of manipulating the probability of different events in various ways. As new information about events becomes available, we can revise the old information and make it more useful. This revised information can sometimes lead to surprising results—as you'll soon see.

This chapter will also teach you how to count. This type of counting, however, goes far beyond what you've seen on *Sesame Street.* Counting events is an important step in calculating probabilities and must be done with care.

In This Chapter

- Using the addition rule of probability
- Using the multiplication rule of probability
- Calculating conditional probabilities
- Using the Bayes' theorem to calculate conditional probabilities
- Using the fundamental counting principle
- Distinguishing between permutations and combinations

Addition Rule of Probabilities

In the previous chapter, we saw the union and intersection of events. Now we are going to put them to work. The *addition rule of probability* calculates the probability of the union of events—that is, the probability that either Event A or Event B will occur. For the addition rule, we need to distinguish between both mutually exclusive events and non-mutually exclusive events (we looked at mutually exclusive events in Chapter 6). The addition rule of probability states the following:

P(A or B) = P(A) + P(B)	If A and B are mutually exclusive events
P(A or B) = P(A) + P(B) – P(A and B)	If A and B are not mutually exclusive events

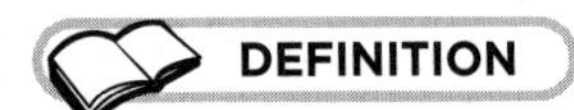

For mutually exclusive events, the **addition rule** states that P(A or B) = P(A) + P(B). If the events are not mutually exclusive, the addition rule states that P(A or B) = P(A) + P(B) - P(A and B).

Now, let's apply these two rules to examples so we can see the difference between them. In rolling a single die, what is the probability of getting a 3 or a 4?

First stop and ask yourself if you can get a 3 and a 4 at the same time when you roll a single die. If the answer is no, like in this example, then the events are mutually exclusive and you don't need to include the third term of the equation. So the probability of rolling a 3 or a 4 is:

$$P(3 \text{ or } 4) = P(3) + P(4) = \frac{1}{6} + \frac{1}{6} = \frac{2}{6} = 0.333$$

In another example, if you draw a card from a deck, what is the probability of getting a heart (H) or a jack (J)? P(H or J) = P(H) + P(J). Stop and ask yourself if you can draw a heart and a jack at the same time. If the answer is yes, like in this example, then the events are not mutually exclusive and you need to include the third term of the equation. There are four jacks in the deck of cards, and one of them is a heart, so this card has been counted twice: once as a jack (included in the four jacks) and once as a heart (included in the 13 hearts). Therefore we need to subtract it once to avoid double counting the card. The probability of getting a heart or a jack is:

$$P(H \text{ or } J) = P(H) + P(J) - P(H \text{ and } J) = \frac{13}{52} + \frac{4}{52} - \frac{1}{52} = \frac{16}{52} = 0.3077$$

To make sure we understand this rule, let's apply it to one more example. The following contingency table represents the grade and the gender for 20 students in a statistics class.

Grade Range	M	F	Total
51-60	1	2	3
61-70	1	2	3
71-80	4	1	5
81-90	3	2	5
91-100	1	3	4
Total	10	10	20

If I want to know the probability that a randomly selected student in class has a grade in the 80s or 90s, I can use the addition rule: P(80s or 90s) = P(80s) + P(90s). I know I should not include the third term since these are mutually exclusive events. So the probability is:

$$P(80s\text{ or }90s) = P(80s) + P(90s) = \frac{5}{20} + \frac{4}{20} = \frac{9}{20} = 0.45$$

If I want to know the probability that a randomly selected student in class is a male or has a grade in the 80s, I can use the addition rule again: P(M or 80s) = P(M) + P(80s). This time, I know I should also include the third term since these are not mutually exclusive events. So the probability is:

$$P(M\text{ or }80s) = P(M) + P(80s) - P(M\text{ and }80s) = \frac{10}{20} + \frac{5}{20} - \frac{3}{20} = \frac{12}{20} = 0.60$$

As you can see from the table, there are 3 students who are male with grades in the 80s. These 3 students are counted twice: once because they are male (included in the 10 males) and once because they have grades in the 80s (included in the 5 students with grades in the 80s). If I didn't subtract these three students, then I would have double counted them.

Before we look at the multiplication rules of probability, we need to know another concept—conditional probability.

Conditional Probability

Conditional probability is the probability that Event A occurs given that (or on the *condition* that) Event B has already occurred. In other words, the probability of Event A is affected by the fact that Event B has already occurred. This means that the two events, A and B, are not independent of each other (Do you remember the independent events we talked about in the last chapter? It all comes back!). We use a different notation for conditional probability—the probability of A given B is written as P(A|B).

To demonstrate, in rolling a single die, what is the probability of getting a 4 given that it is an even number? Now see the difference in wording! If I say the probability of getting a 4 without any conditions, we know it is $\frac{1}{6}$. However, I added a condition. (Let's say I peeked and gave you the additional information that the number I rolled was an even number.) This condition limits your sample space to only 3 instead of 6. Therefore, $P(4|\text{Even}) = \frac{1}{3} = 0.3333$.

Let's look at another example to make sure we master the conditional probability concept. In drawing a card from a deck of cards, what is the probability that it is a jack given that it is a face card? Now we know the trick—I limited your sample space to the condition given, which is a face card. So the sample space is now 12. Then the probability of drawing a jack (J) given that it is a face (F) card is $P(J|F) = \frac{4}{12} = 0.3333$.

As you might imagine, there is a formula to calculate the conditional probability:

$$P(A|B) = \frac{P(A \text{ and } B)}{P(B)}$$

Let's apply this formula to our grade and gender example. If I want to know the probability that a randomly selected student has a grade in the 80s given that she is a female, I can plug it in as follows:

$$P(80s|F) = \frac{P(80s \text{ and } F)}{P(F)} = \frac{2/20}{10/20} = \frac{2}{10} = 0.20$$

Now just for fun, I'll introduce you to some more statistics jargon: prior probability and posterior probability. The probability of rolling a 4 without any condition is called a simple probability, or a *prior* probability, because it is derived only from the information currently available (that is, prior to receiving any conditional information). On the other hand, conditional probability is also known as *posterior* probability because it's based on a revision of the prior probability, given additional information.

Simple or **prior** probabilities are always based on the total number of observations. Conditional or **posterior** probabilities are based on the probability of Event A given the information that Event B has already occurred.

Conditional probability is very useful for determining the probability of compound events, as you will see in the following section.

Multiplication Rule of Probabilities

The *multiplication rule of probability* calculates the probability of the intersection of events, that is, the probability that both Event A and Event B will occur. The probability of Event A and Event B occurring at the same time is also called a *joint probability*. For the multiplication rule, we need to distinguish between independent and dependent events. The multiplication rule of probability states the following:

$P(A \text{ and } B) = P(A) \times P(B)$ If A and B are independent events

$P(A \text{ and } B) = P(A) \times P(B|A)$ If A and B are dependent events

DEFINITION

For independent events, the **multiplication rule** states that P(A and B) = P(A) × P(B). If the events are dependent, the multiplication rule states that P(A and B) = P(A) × P(B|A).

To demonstrate using independent events, in flipping a coin two times, what is the probability of getting a head the first time (H_1) and a head the second time (H_2)? Since these are independent events, we use the following formula: $P(H_1 \text{ and } H_2) = P(H_1) \times P(H_2) = \frac{1}{2} \times \frac{1}{2} = \frac{1}{4} = 0.25$

To demonstrate the multiplication rule with dependent events, let's go back to our grade and gender example. If I want to know the probability that a randomly selected student is female (F) with a grade in the 80s, then we use the following formula: $P(F \text{ and } 80s) = P(F) \times P(80s|F) = \frac{10}{20} \times \frac{2}{10} = \frac{2}{20} = 0.10$

When you have the data in a contingency table, it is easier to get the probability of F and 80s—just locate the number of females with grades in the 80s from the table (2), and then divide it by the total number of observations (20). Let's prove the point by applying it to another example. What is the probability that a randomly selected student is male with a grade in the 90s? If you replied $\frac{1}{20}$, then yes, you are correct. I agree, this method is way easier!

BOB'S BASICS

The *addition rule* of probability calculates the probability that Event A *or* Event B occurs, whereas the *multiplication rule* of probability calculates the probability that Event A *and* Event B occur. To remember it, the **addition rule** calculates **OR** while the **multiplication rule** calculates **AND**.

Since you now understand the probability rules of addition and multiplication, I'm going to introduce you next to the Bayes' theorem, which is conditional probability in a more general format.

Bayes' Theorem

Thomas Bayes (1701–1761) developed a mathematical rule that deals with calculating $P(A|B)$ from the information about $P(B|A)$. Bayes' theorem states the following:

$$P(A|B) = \frac{P(A)P(B|A)}{P(A)P(B|A)+P(A')P(B|A')}$$

Where:

$P(A')$ = the probability of the complement of Event A

$P(B|A')$ = the probability of Event B, given that the complement to Event A has occurred.

Now that looks like a mouthful, but applying it to a new example will clear things up. In a different statistics class, 45 percent of students are male with a 75 percent probability of passing. The female students have an 80 percent probability of passing. If I randomly pick a student who is passing (PS), what is the probability that this student is female?

Students always say the most difficult part of Bayes' theorem is all of the different parts, so let's clear this up a little and break down the example. The pieces given in the example are:

$P(M) = 0.45$

$P(PS|M) = 0.75$

$P(PS|F) = 0.8$

You can easily figure that 55 percent of the students are female. So to calculate $P(F|PS)$, we can apply Bayes' theorem as follows:

$$P(F|PS) = \frac{P(F)P(PS|F)}{P(F)P(PS|F)+P(M)P(PS|M)} = \frac{(0.55)(0.8)}{(0.55)(0.8)+(0.45)(0.75)} = \frac{0.44}{0.7775} = 0.57$$

Knowing that the student passed the exam, there is a 57 percent chance that the student is female. See, Bayes' theorem is not so difficult after all!

RANDOM THOUGHTS

Not only was Thomas Bayes a prominent mathematician, but he was also a published Presbyterian minister who used mathematics to study religion.

You've learned the probability rules, so let's move on to the last topic of this chapter, the counting principles.

Counting Principles

As we learned in the last chapter, to be able to use the classical probability approach, you need to know the total number of outcomes possible for the event of interest (the sample space). For simple events like rolling a die or drawing a card from a deck, it's easy to count the number of outcomes in the sample space. But for other events like a state lottery drawing, finding the total number of outcomes is difficult. In these cases, counting rules come in handy to help us determine the number of possible outcomes of an experiment or event.

There are three main counting rules: the fundamental counting rule, permutations, and combinations. Let's look at each.

The Fundamental Counting Rule

On a nice Sunday afternoon, your mom took you out for an ice cream. The ice cream shop has many choices, and you are standing there for some time trying to determine which to choose. There are eight flavors (vanilla, chocolate, mango, strawberry, peach, banana, raspberry, and coffee), two toppings (hot fudge and sprinkles), and three sizes (small, medium, and large). While you are looking at the assortment, you wonder how many different choices you have. The *fundamental counting rule* can help. This rule says that if one event (my flavor choice) can occur in m ways, and a second event (my topping choice) can occur in n ways, then the total number of possible ways that these two events can occur together is $m \times n$ ways. You can stretch this principle to more than two events.

If I extend this rule to all three choices I have in the ice cream shop, then to get the total number of choices I use multiplication: 8 flavors × 2 toppings × 3 sizes = 48 choices. When your mom says you are taking too long to decide, you tell her that you have 48 ice cream choices to choose from, and they're summarized in the table below. Better make up your mind soon!

Ice Cream Combinations (Flavor, Topping, Size)			
Vanilla, hot fudge, small	Mango, hot fudge, small	Peach, hot fudge, small	Raspberry, hot fudge, small
Vanilla, hot fudge, medium	Mango, hot fudge, medium	Peach, hot fudge, medium	Raspberry, hot fudge, medium
Vanilla, hot fudge, large	Mango, hot fudge, large	Peach, hot fudge, large	Raspberry, hot fudge, large
Vanilla, sprinkle, small	Mango, sprinkle, small	Peach, sprinkle, small	Raspberry, sprinkle, small
Vanilla, sprinkle, medium	Mango, sprinkle, medium	Peach, sprinkle, medium	Raspberry, sprinkle, medium

continues

continued

Ice Cream Combinations (Flavor, Topping, Size)			
Vanilla, sprinkle, large	Mango, sprinkle, large	Peach, sprinkle, large	Raspberry, sprinkle, large
Chocolate, hot fudge, small	Strawberry, hot fudge, small	Banana, hot fudge, small	Coffee, hot fudge, small
Chocolate, hot fudge, medium	Strawberry, hot fudge, medium	Banana, hot fudge, medium	Coffee, hot fudge, medium
Chocolate, hot fudge, large	Strawberry, hot fudge, large	Banana, hot fudge, large	Coffee, hot fudge, large
Chocolate, sprinkle, small	Strawberry, sprinkle, small	Banana, sprinkle, small	Coffee, sprinkle, small
Chocolate, sprinkle, medium	Strawberry, sprinkle, medium	Banana, sprinkle, medium	Coffee, sprinkle, medium
Chocolate, sprinkle, large	Strawberry, sprinkle, large	Banana, sprinkle, large	Coffee, sprinkle, large

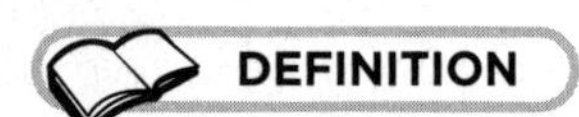

The **fundamental counting rule** gives the total number of possible ways for multiple events to occur together. If the first event can occur in m ways and the second event can occur in n ways, then the total number of ways both events can occur together is $m \times n$ ways. You can extend this rule to more than two events.

Another example of the fundamental counting rule is the number of automobile license plates that a state can issue. Pennsylvania license plates have three letters followed by four numbers. Your friend, who has not taken a statistics class yet, is telling you if the state eliminates the use of the letter O and the number 0 from license plates because they look alike, then the total number of possible license plates that Pennsylvania can issue will not decrease by much (because it's only two figures). However, you have studied the counting rules and you know this is incorrect. To prove it to your friend, you showed him the total number of possible license plates Pennsylvania can issue in both cases as follows:

If all letters and numbers are allowed, using the fundamental counting rule, then the total number of possible license plate combinations is $26 \times 26 \times 26 \times 10 \times 10 \times 10 \times 10 = 175{,}760{,}000$ combinations. If Pennsylvania eliminates the letter O and the number 0 to avoid confusion

since they look alike, the total number of possible license plates is
$25 \times 25 \times 25 \times 9 \times 9 \times 9 \times 9 = 102{,}515{,}625$ choices. You proved your point to your friend and he was impressed by your knowledge, so he decided to take a statistics class next semester!

TEST YOUR KNOWLEDGE

Your younger sister is fascinated with states' license plates. She noticed that Maryland license plates have three letters followed by three numbers, while New Jersey license plates have four letters and two numbers. Since each states' license plates have combinations of six figures on them, she thinks that Maryland and New Jersey can issue the same number of possible license plates. She asks you if this is right, and you tell her "no." Maryland can issue 17,576,000 plates ($26 \times 26 \times 26 \times 10 \times 10 \times 10$), while New Jersey can issue 45,697,600 plates ($26 \times 26 \times 26 \times 26 \times 10 \times 10$). She's very surprised by your answer and says, "I'm glad I asked you!"

When you use the fundamental counting rule to determine the total number of possible outcomes, it's important to know whether or not repetition is allowed. The previous rule assumes repetition is allowed, but if repetition is not allowed, then the number of ways in which each event can occur will decrease. Let's see the difference between the two rules. In your summer job in a local grocery store, the manager asks you to design ID cards for employees. He wants each ID card to have four digits on it. He asks you how many ID cards he can make if repetition is allowed and if repetition is not allowed. You tell him that if repetition is allowed, then you can have 10,000 ID cards, but if repetition isn't allowed, then you can only have 5,040 ID cards. You explain:

> If repetition is allowed, then each digit on the ID card has 10 choices. So the first digit can be chosen out of 10 numbers, the second digit can be chosen out of 10 numbers, and so can the third and fourth digits. This gives us a total of $10 \times 10 \times 10 \times 10 = 10{,}000$ ID cards.
>
> If repetition isn't allowed (meaning the same number can't be used more than once on each ID card), then the first digit on the ID card can be chosen out of 10 digits, but the second digit can only be chosen out of 9 digits—since one digit has already been chosen, it can't be repeated. Then the third digit on the ID card can only be chosen out of 8 digits, and so on. Without repetition, the total number of ID cards we can make now is $10 \times 9 \times 8 \times 7 = 5{,}040$ ID cards.

The manager says he's glad he asked you!

Permutations

Permutations are the number of different ways in which r objects can be chosen one at a time from n distinct objects, where the order is important. It can be found with ${}_nP_r = \frac{n!}{(n-r)!}$.

DEFINITION

Permutations are the number of different ways in which objects can be arranged in order. The number of permutations of n objects taken r at a time can be found by ${}_nP_r \frac{n!}{(n-r)!}$.

BOB'S BASICS

6 × 5 × 4 × 3 × 2 × 1 is written as "6!" and is pronounced "6 factorial." In general, $n! = n \times (n-1) \times (n-2) \times (n-3) \ldots 4 \times 3 \times 2 \times 1$.

For example, a TV director for the evening news wants to use 5 new stories to broadcast on the evening news. One story will be broadcast first, one will be second, one will be third, one will be fourth, and the last one will be the closing story. If the TV director has 10 new stories to choose from, in how many possible ways can the evening news show be set up?

Since the order of the stories is important, we use permutations. The number of possible ways the evening news show can be set up is

${}_{10}P_5 = \frac{10!}{(10-5)!} = \frac{(10)(9)(8)(7)(6)(5!)}{5!} = (10)(9)(8)(7)(6) = 30{,}240$ different ways to set up the evening news show. I'm glad I'm not the TV evening news director!

BOB'S BASICS

It is easier to calculate the number of permutations using this formula:

$${}_nP_r = \frac{n!}{(n-r)!} = \frac{n\ (n-1)\ (n-2)\ldots(n-r+1)\ (n-r)!}{(n-r)!} = n \cdot (n-1) \cdot (n-2) \ldots (n-r+1).$$

This is because $(n-r)!$ in the numerator cancels out the denominator.

TEST YOUR KNOWLEDGE

In 1808, mathematician Christian Kramp was the first to introduce the factorial notation.

Combinations

Combinations are similar to permutations except that the order of the objects is not important. The number of combinations in which r objects are selected from n possible objects can be found as follows:

$${}_nC_r = \frac{n!}{r!(n-r)!}$$

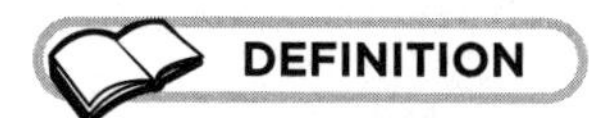

Combinations are the number of different ways in which objects can be arranged without regard to the order. The number of combinations of n objects taken r at a time can be found by ${}_nC_r = \frac{n!}{r!(n-r)!}$.

The number of combinations is always less than the number of permutations. In the combinations formula, dividing by $r!$ removes the duplicates from the number of permutations. For each two letters, there are two permutations (since order is important, so AB is different from BA) but only one combination (since order is not important, so AB and BA are the same). Thus, dividing the permutation by $r!$ eliminates the duplicates. To see that clearly, let's look at the following example.

From the first 4 letters of the alphabet (A, B, C, and D), how many permutations of 2 letters can we get? How many combinations of 2 letters can we get?

$$\text{Number of permutations } {}_4P_2 = \frac{4!}{(4-2)!} = \frac{(4)(3)(2!)}{2!} = (4)(3) = 12$$

$$\text{Number of combinations } {}_4C_2 = \frac{4!}{2!(4-2)!} = \frac{(4)(3)(2!)}{2!(2!)} = \frac{(4)(3)}{(2)(1)} = 6$$

To see the difference, the permutations are:

AB	BA	CA	DA
AC	BC	CB	DB
AD	BD	CD	DC

Whereas the combinations are:

AB			
AC	BC		
AD	BD	CD	

By eliminating the duplicates from the permutations, we get the combinations.

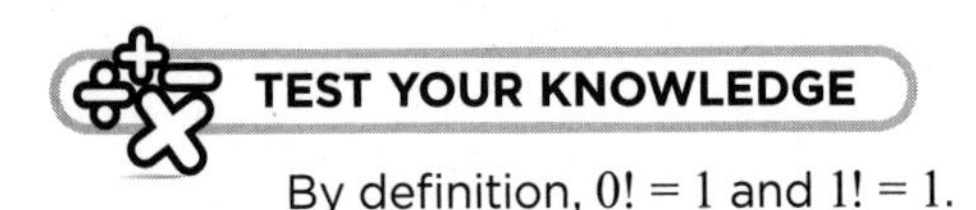

TEST YOUR KNOWLEDGE

By definition, $0! = 1$ and $1! = 1$.

Combinations are often used in the selection of committees. As an example, your tennis club wants to choose an executive committee of 5 people. There are 12 people from whom to choose. How many different possibilities are there?

Number of combinations ${}_{12}C_5 = \frac{12!}{5!(12-5)!} = \frac{(12)(11)(10)(9)(8)(7!)}{(5)(4)(3)(2)(1)(7!)} = \frac{(12)(11)(10)(9)(8)}{(5)(4)(3)(2)(1)} = 792$ different ways to choose the committee.

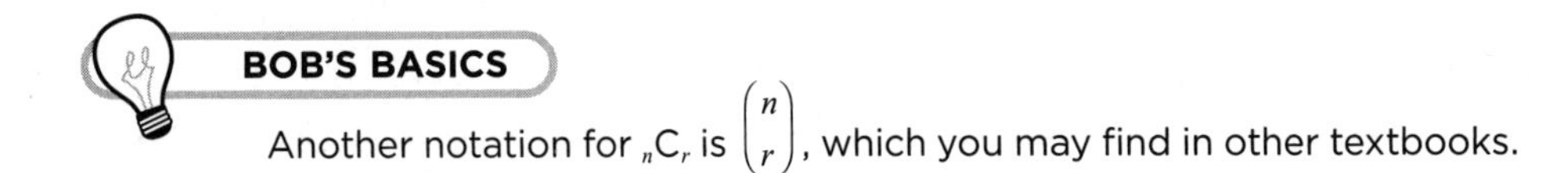

BOB'S BASICS

Another notation for ${}_nC_r$ is $\binom{n}{r}$, which you may find in other textbooks.

Using Excel to Calculate Factorials, Permutations, and Combinations

Excel has built-in statistical functions that help us in all sorts of calculations, including factorials, permutations, and combinations. The Excel functions are:

Factorial = FACT(n)

Function Arguments ? ×

FACT

Number 6 = 6

= 720

Returns the factorial of a number, equal to 1*2*3*...* Number.

Number is the nonnegative number you want the factorial of.

Formula result = 720

Help on this function OK Cancel

Figure 7.1

Factorial function in Excel.

Permutations = PERMUT(n,r)

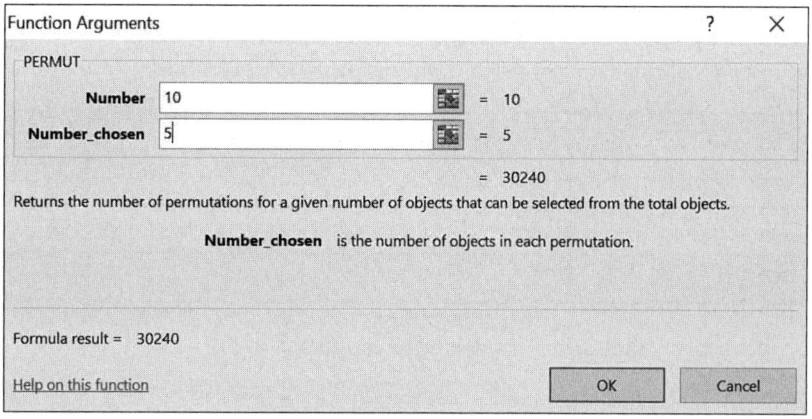

Function Arguments ? X

PERMUT

Number 10 = 10

Number_chosen 5 = 5

= 30240

Returns the number of permutations for a given number of objects that can be selected from the total objects.

Number_chosen is the number of objects in each permutation.

Formula result = 30240

Help on this function OK Cancel

Figure 7.2

Permutations function in Excel.

Combinations = COMBIN(n,r)

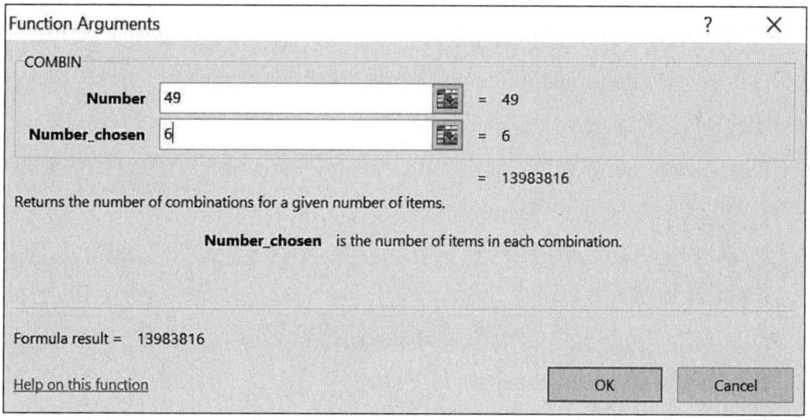

Function Arguments ? X

COMBIN

Number 49 = 49

Number_chosen 6 = 6

= 13983816

Returns the number of combinations for a given number of items.

Number_chosen is the number of items in each combination.

Formula result = 13983816

Help on this function OK Cancel

Figure 7.3

Combinations function in Excel.

This concludes our discussion of probability and counting rules. Test your understanding by doing the practice problems.

Practice Problems

1. A political telephone survey asked 260 people whether they were in favor or not in favor of a proposed law. Each person was identified as a Republican or Democrat. The following contingency table shows the results.

Party	In Favor	Not in Favor	Total
Republican	98	54	152
Democrat	79	29	108
Total	177	83	260

A person from the survey is selected at random. We define:

Event A: The person selected is in favor of the new law.

Event B: The person selected is a Republican.

a. Determine the probability that the selected person is in favor of the new law.

b. Determine the probability that the selected person is a Republican.

c. Determine the probability that the selected person is not in favor of the new law.

d. Determine the probability that the selected person is a Democrat.

e. Determine the probability that the selected person is in favor of the new law given that the person is a Republican.

f. Determine the probability that the selected person is not in favor of the new law given that the person is a Republican.

g. Determine the probability that the selected person is in favor of the new law given that the person is a Democrat.

h. Determine the probability that the selected person is in favor of the new law and that the person is a Republican.

i. Determine the probability that the selected person is in favor of the new law and that the person is a Democrat.

j. Determine the probability that the selected person is in favor of the new law or that the person is a Republican.

k. Determine the probability that the selected person is in favor of the new law or that the person is a Democrat.

l. Using Bayes' theorem, calculate the probability that the selected person was a Republican, given that the person was in favor of the new law.

2. A survey asked 125 families whether the household had Internet access. Each family was classified by race. The following contingency table shows the results.

Race	**Internet**	**No Internet**	**Total**
Caucasian	15	22	37
Asian American	23	18	41
African American	14	33	47
Total	52	73	125

   A family from the survey is randomly selected. We define:

   Event A: The selected family has an Internet connection in its home.

   Event B: The selected family is Asian American.

   a. Determine the probability that the selected family has an Internet connection.

   b. Determine the probability that the selected family is Asian American.

   c. Determine the probability that the selected family has an Internet connection and is Asian American.

   d. Determine the probability that the selected family has an Internet connection or is Asian American.

3. A restaurant has a menu with three appetizers, eight entrées, four desserts, and three drinks. How many different meals can you order?

4. A multiple-choice test has 10 questions, with each question having four choices. What is the probability that a student, who randomly answers each question, will answer each question correctly?

5. The NBA teams with the 13 worst records at the end of the season participate in a lottery to determine the order in which they will draft new players for the next season. How many different arrangements exist for the drafting order for these 13 teams?

6. In a race with eight swimmers, how many ways can swimmers finish first, second, and third?

7. How many different ways can 10 new movies be ranked first and second by a movie critic?

8. A combination lock has a total of 40 numbers and will unlock with the proper three-number sequence. How many possible combinations exist?

9. I would like to select three paperback books from a list of 12 books to take on vacation. How many different sets of three books can I choose?

10. A panel of 12 jurors needs to be selected from a group of 50 people. How many different juries can be selected?

The Least You Need to Know

- For mutually exclusive events, the addition rule states that P(A or B) = P(A) + P(B). If the events are not mutually exclusive, the addition rule becomes P(A or B) = P(A) + P(B) – P(A and B).
- We define conditional probability as the probability of Event A knowing that Event B has already occurred.
- For independent events, the multiplication rule states that P(A and B) = P(A) P(B). If the events are dependent, the multiplication rule becomes P(A and B) = P(A) P(B|A).
- Bayes' theorem deals with calculating P(A|B) from information about P(B|A) using the following formula:

  $$\frac{P(A)P(B|A)}{P(A)P(B|A)+P(A')P(B|A')}$$

- The fundamental counting principle states that if one event can occur in *m* ways and a second event can occur in *n* ways, then the total number of ways both events can occur together is $m \cdot n$ ways. We can extend this principle to more than two events.
- Permutations are the number of different ways in which objects can be arranged in order. Combinations are the number of different ways in which objects can be arranged when order is of no importance.

CHAPTER

8

Probability Distributions

In This Chapter

- Defining a random variable and probability distribution
- Calculating the mean and variance of a discrete probability distribution

Well, we've finally arrived at our third and last chapter on general probability concepts. This chapter sets the stage for the last three chapters in Part 2, which will focus on specific types of probability distributions. Before you know it, we'll be knee deep with inferential statistics.

In this chapter, we will look at the basic concepts of probability distributions and then move on to calculate the mean and variance of probability distributions.

Basic Concepts of Probability Distributions

Now let's introduce you to probability distributions and prepare you for the last three chapters of Part 2. However, first we need to discuss the topic of random variables, which will lay the groundwork for specific probability distributions in Chapters 9, 10, and 11.

Random Variables

Random variables are variables whose values cannot be determined with certainty before conducting the experiment. They can take on different values and result from experiments. Examples of experiments could be rolling a single die, weighing three randomly selected candy bars, or measuring the wait time for customers in a checkout line.

Random variables can be discrete or continuous. (Does this ring a bell? We talked about the difference in Chapter 2.) A *discrete random variable* can assume only whole numbers and results from counting the outcomes of an experiment. Discrete random variables come from experiments like counting the number of customers waiting in the checkout line or the outcomes of flipping a coin or rolling a die. A *continuous random variable*, on the other hand, can assume any numerical value within an interval and results from measuring the outcome of an experiment. Examples of continuous random variables are the wait time of customers in a checkout line, the weight of your favorite candy bar, or the time it takes to fly from Philadelphia to Chicago.

A **random variable** is an outcome that takes on a numerical value as a result of an experiment. Random variables can be discrete or continuous. A **discrete random variable** can assume only whole numbers and results from *counting* the outcomes of an experiment. A **continuous random variable** can assume any numerical value within an interval and results from *measuring* the outcome of an experiment.

Probability Distributions

Random variables have *probability distributions* associated with them. A probability distribution is nothing more than a table with two columns: the values of the random variable in one column and the probability associated with each value in the second column. For example, in rolling a single die, we know the outcomes and the probability of getting each one. I can put them in a table as follows:

Outcome	Probability
1	$\frac{1}{6} = 0.167$
2	$\frac{1}{6} = 0.167$
3	$\frac{1}{6} = 0.167$
4	$\frac{1}{6} = 0.167$
5	$\frac{1}{6} = 0.167$
6	$\frac{1}{6} = 0.167$

This is the probability distribution. Simple, isn't it?

A **probability distribution** is a table that lists all the values of the random variable and the probability associated with each value.

A probability distribution can be discrete or continuous, depending on the type of random variable used. If the random variable is discrete, then the probability distribution is called a discrete probability distribution. If the random variable is continuous, then the probability distribution is called a continuous probability distribution. We will see examples of each type in the next three chapters. I know it sounds simple—statistics is not difficult, after all!

Chapters 9 and 10 give examples of common discrete probability distributions, whereas Chapter 11 gives an example of a very familiar continuous probability distribution. For now, we will focus solely on discrete random variables.

Let's look at another example. Bob's oldest daughter, Christin, was a very accomplished competitive swimmer between the ages of 7 and 13, but her talent certainly didn't come from Bob's side of the family. Christin could not only swim, but she also could swim *fast.*

The following table is a relative frequency distribution showing the number of first-, second-, third-, fourth-, and fifth-place finishes that Christin earned during 50 races.

Place	Number of Races	Relative Frequency (Probability)
1	27	$\frac{27}{50} = 0.54$
2	12	$\frac{12}{50} = 0.24$
3	7	$\frac{7}{50} = 0.14$
4	3	$\frac{3}{50} = 0.06$
5	1	$\frac{1}{50} = 0.02$
	Total = 50	= 1.00

If we define the random variable x = the place Christin finished in a race, the previous table would be the discrete probability distribution for the variable x. From this table, we can state the probability that Christin will finish first as follows:

$$P(x = 1) = 0.54$$

Figure 8.1 shows the discrete probability distribution for x graphically.

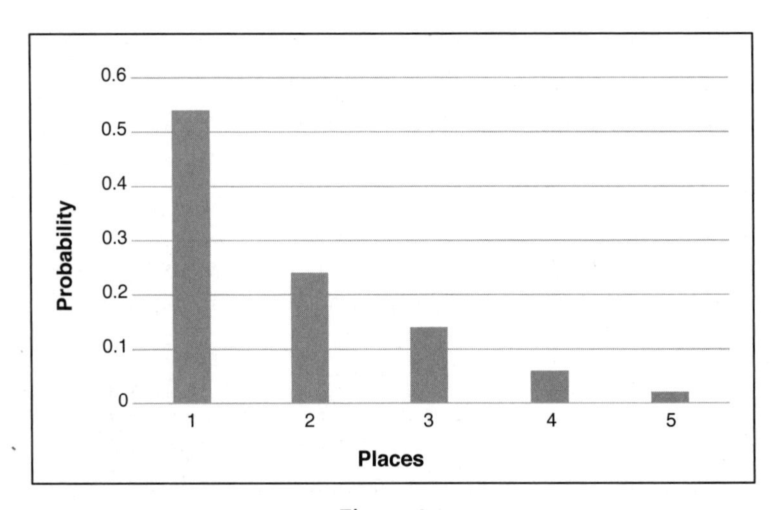

Figure 8.1
The discrete probability distribution for Christin's races.

Rules for Discrete Probability Distributions

Any discrete probability distribution needs to meet the following requirements:

- Each outcome in the distribution needs to be mutually exclusive—that is, the value of the random variable cannot fall into more than one of the frequency distribution classes. For example, it is not possible for Christin to take first and second place in the same race.
- The probability of each outcome, $P(x)$, must be between 0 and 1; that is, $0 \leq P(x) \leq 1$ for all values of x. In the previous example, $P(x = 3) = 0.14$, which falls between 0 and 1.
- The sum of the probabilities for all the outcomes in the distribution needs to add up to 1; that is, $\sum_{i=1}^{n} P(x_i) = 1$. In the swimming example, the sum of the Relative Frequency (Probability) column in the previous table adds up to 1.

The Mean of Discrete Probability Distributions

The mean of a discrete probability distribution is simply a weighted average (discussed in Chapter 4) calculated using the following formula:

$$\mu = \sum_{i=1}^{n} x_i P(x_i)$$

where:

μ = the mean of the discrete probability distribution

x_i = the value of the random variable for the *i*th outcome

$P(x_i)$ = the probability that the *i*th outcome will occur

n = the number of outcomes in the distribution

The table that follows revisits Christin's swimming probability distribution.

Place (x_i)	**Probability** $P(x_i)$
1	0.54
2	0.24
3	0.14
4	0.06
5	0.02

The mean of this discrete probability distribution is as follows:

$$\mu = \sum_{i=1}^{5} x_i P(x_i)$$

$$\mu = (1)(0.54) + (2)(0.24) + (3)(0.14) + (4)(0.06) + (5)(0.02) = 1.78$$

This mean is telling us that Christin's average finish for a race is 1.78 place! How does she do that? Obviously, this will never be the result of any one particular race. Rather, it represents the average finish of many races. The mean of a discrete probability distribution does not have to equal one of the values of the random variable (1, 2, 3, 4, or 5 in this case).

Another term for describing the mean of a probability distribution is the *expected value*, E(x). Therefore:

$$E(x) = \mu = \sum_{i=1}^{n} x_i P(x_i)$$

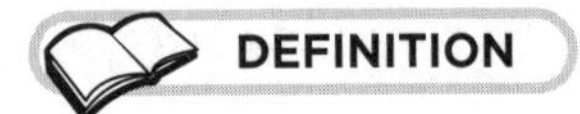

DEFINITION

An **expected value** is the mean of a probability distribution.

Didn't I say statisticians love all sorts of notation to describe the same concept?

The Variance and Standard Deviation of Discrete Probability Distributions

Just when you thought it was safe to get back into the water, along comes another variance! Well, if you've seen one variance calculation, you've seen them all. You can calculate the variance for a discrete probability distribution as follows:

$$\sigma^2 = \sum_{i=1}^{n} (x_i - \mu)^2 P(x_i)$$

where:

σ^2 = the variance of the discrete probability distribution

As before, the standard deviation of the distribution is as follows:

$$\sigma = \sqrt{\sigma^2}$$

To demonstrate the use of these equations, we'll rely on Christin's swimming distribution. The calculations are summarized in the following table.

x_i	$P(x_i)$	μ	$x_i - \mu$	$(x_i - \mu)^2$	$(x_i - \mu)^2 P(x_i)$
1	0.54	1.78	-0.78	0.608	0.328
2	0.24	1.78	0.22	0.048	0.012
3	0.14	1.78	1.22	1.488	0.208
4	0.06	1.78	2.22	4.928	0.296
5	0.02	1.78	3.22	10.368	0.208

$$\sigma^2 = \sum_{i=1}^{5} (x_i - \mu)^2 P(x_i) = 0.328 + 0.12 + 0.208 + 0.296 + 0.208 = 1.052$$

The standard deviation of this distribution is:

$$\sigma = \sqrt{\sigma^2} = \sqrt{1.052} = 1.026$$

A more efficient way to calculate the variance of a discrete probability distribution is:

$$\sigma^2 = \left(\sum_{i=1}^{n} x_i^2 P(x_i)\right) - \mu^2$$

The following table summarizes these calculations using Christin's swimming example.

x_i	$P(x_i)$	x_i^2	$x_i^2\ P(x_i)$
1	0.54	1	0.54
2	0.24	4	0.96
3	0.14	9	1.26
4	0.06	16	0.96
5	0.02	25	0.50

$$\sigma^2 = \left(\sum_{i=1}^{n} x_i^2 P(x_i)\right) - \mu^2$$

$$\sigma^2 = (0.54 + 0.96 + 1.26 + 0.96 + 0.50) - (1.78)^2$$

As you can see, the result is the same, but with less effort!

Practice Problems

1. Stock "A" has the following probability distribution for its rate of return:

Rate of Return for Stock A (R_A)	**Probability**
0.1	0.3
0.15	0.5
0.1	0.2

Calculate the mean, variance, and standard deviation for the rate of return on stock "A."

2. A survey of 450 families was conducted to find how many cats were owned by each respondent. The following table summarizes the results.

Number of Cats	Number of Families
0	137
1	160
2	112
3	31
4	10

Develop a probability distribution for this data and calculate the mean, variance, and standard deviation.

The Least You Need to Know

- A random variable is an outcome that takes on a numerical value as a result of an experiment. The value is not known with certainty before the experiment.
- A random variable is discrete if it is limited to assuming only specific integer values as a result of counting the outcome of an experiment. A random variable is continuous if it can assume any numerical value within an interval as a result of measuring the outcome of an experiment.
- A probability distribution is a listing of all the possible outcomes of an experiment along with the relative frequency or probability of each outcome.
- You find the mean, or expected value, of a discrete probability distribution as follows:

$$E(x) = \mu = \sum_{i=1}^{n} x_i P(x_i) \cdot$$

- You find the variance of a discrete probability distribution as follows:

$$\sigma^2 = \sum_{i=1}^{n} (x_i - \mu)^2 P(x_i) \cdot$$

CHAPTER

9

The Binomial Probability Distribution

Our discussion of discrete probability distributions so far has been limited to general distributions based on historical data that has been previously collected. However, some theoretical probability distributions are based on a mathematical formula rather than historical data. We will address the first of these, the binomial probability distribution, in this chapter.

In many types of problems we are interested in the probability of an event occurring several times. A classical example that has been torturing students for many years is "What is the chance of getting 7 heads when tossing a coin 10 times?" By the time you finish this chapter, answering this question will be a piece of cake!

In This Chapter

- Describe the characteristics of a binomial experiment
- Calculate the probabilities for a binomial distribution
- Find probabilities using a binomial table
- Find binomial probabilities using Excel
- Calculate the mean and standard deviation of a binomial distribution

Characteristics of a Binomial Experiment

If you remember, in Chapter 6 we defined experimenting as the process of measuring or observing an activity for the purpose of collecting data. Let's say our experiment of interest involves a certain professional basketball player shooting free throws. Each free throw would be considered a *trial* for the experiment. For this particular experiment, we have only two possible outcomes for each trial; either the free throw goes into the basket (a success) or it doesn't (a failure). Because we can have only two possible outcomes for each trial, this is known as a *binomial experiment.*

A **binomial experiment** has the following characteristics: (1) the experiment consists of a fixed number of trials denoted by n; (2) each trial has only two possible outcomes, a success or a failure; (3) the probability of success and the probability of failure are constant throughout the experiment; (4) each trial is independent of any other trial in the experiment.

Let's say that our player of interest is Michael Jordan, who historically has made 80 percent of his free throws. So the probability of success, p, of any given free throw is 0.80. Because there are only two outcomes possible, the probability of failure for any given free throw, q, is 0.20. For a binomial experiment, the values of p and q must be the same for every trial in the experiment. Because only two outcomes are allowed in a binomial experiment, $p = 1 - q$ always holds true.

BOB'S BASICS

The outcomes in the binomial distribution are classified as either success or failure. The word *success* doesn't necessarily mean a positive outcome. It is the outcome we are interested in. Likewise, the word *failure* doesn't necessarily mean a negative outcome.

Finally, a binomial experiment requires that each trial is independent of any other trial. In other words, the probability of the second free throw being successful is not affected by whether the first free throw was successful. Other examples of binomial experiments include the following:

- Testing whether a part is defective after it has been manufactured
- Observing the number of correct responses in a multiple-choice exam
- Counting the number of American households that have an internet connection

Now that we have defined the ground rules for binomial experiments, we are ready to graduate to calculating binomial probabilities.

The Binomial Probability Distribution

The binomial probability distribution allows us to calculate the probability of a specific number of successes for a certain number of trials. Therefore, the random variable for this distribution would be the number of successes that were observed. To demonstrate a binomial distribution, I will use the following example.

Suppose that the probability of passing an exam is 60 percent, so the probability of failing the exam is 40 percent. This represents a binominal experiment, with $p = 0.60$ (the probability of a "success") and $q = 0.40$ (the probability of a "failure"). We can calculate the probability of x successes in n trials using the binomial distribution, as follows:

$$P(x) = \frac{n!}{x!(n-x)!} p^x q^{(n-x)}$$

Where:

n: the total number of trials

x: the number of successes

p: the probability of a success

q: the probability of failure

$P(x)$: the probability of observing x number of successes in n number of trials.

If the class has five students, with this equation, we can calculate the probability that three students will pass the exam as follows:

$$P(3) = \frac{5!}{3!(5-3)!}(0.6)^3(0.4)^{5-3}$$

$$P(3) = \left(\frac{120}{2 \cdot 6}\right)(0.216)(0.16) = 0.3456$$

There is a 34.56 percent chance that three students out of the class of five students will pass the exam. We can also calculate the probability that zero, one, two, four, or five students will pass the exam as follows:

BOB'S BASICS

Remember from Chapter 7, $0! = 1$. Also $x^0 = 1$ for any value of x.

For $x = 0$:

$$P(0) = \frac{5!}{0!(5-0)!}(0.6)^0(0.4)^{5-0}$$

$$P(0) = \left(\frac{120}{120 \cdot 1}\right)(1)(0.0102) = 0.0102$$

For $x = 1$:

$$P(1) = \frac{5!}{1!(5-1)!}(0.6)^1(0.4)^{5-1}$$

$$P(1) = \left(\frac{120}{24 \cdot 1}\right)(0.6)(0.0256) = 0.0768$$

For $x = 2$:

$$P(2) = \frac{5!}{2!(5-2)!}(0.6)^2(0.4)^{5-2}$$

$$P(2) = \left(\frac{120}{6 \cdot 2}\right)(0.36)(0.064) = 0.2304$$

For $x = 4$:

$$P(4) = \frac{5!}{4!(5-4)!}(0.6)^4(0.4)^{5-4}$$

$$P(4) = \left(\frac{120}{1 \cdot 24}\right)(0.1296)(0.4) = 0.2592$$

For $x = 5$:

$$P(5) = \frac{5!}{5!(5-5)!}(0.6)^5(0.4)^{5-5}$$

$$P(5) = \left(\frac{120}{1 \cdot 120}\right)(0.0778)(1) = 0.0778$$

The following table summarizes all the previous probabilities.

x	P(*x*)
0	0.0102
1	0.0768
2	0.2304
3	0.3456
4	0.2595
5	0.0778
	Total = 1.0

This table represents the binomial probability distribution for *x* successes in five trials with the probability of success equal to 0.60. Notice that the sum of all the probabilities equals 1, which is a requirement for all probability distributions. Figure 9.1 shows this probability distribution as a histogram.

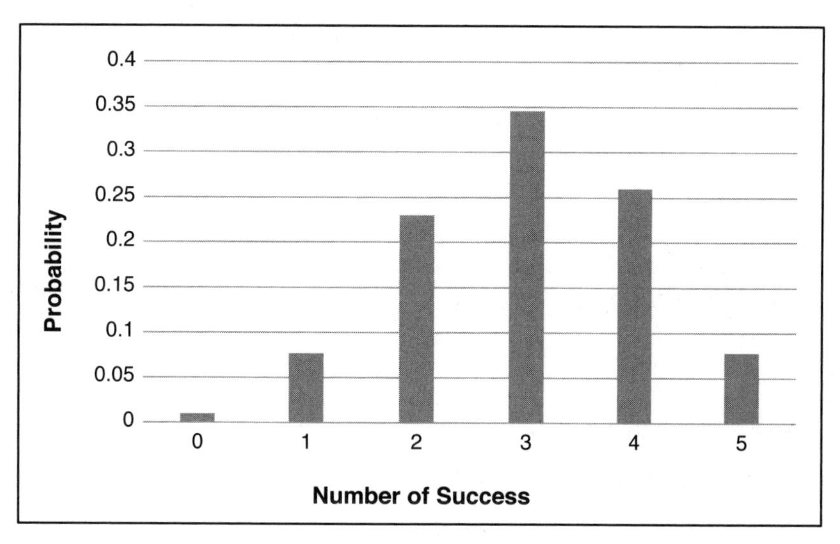

Figure 9.1
Binomial probability distribution.

From this figure, we can see that the most likely number of students who will pass the exam out of the class of 5 students is 3.

Finally, we can calculate the probability of multiple events for this distribution. For instance, the probability that three or more students in the class will pass the exam is this:

$$P(x \geq 3) = P(x = 3) + P(x = 4) + P(x = 5)$$

$$P(x \geq 3) = 0.3456 + 0.2595 + 0.0778 = 0.6829$$

Also, the probability that one or less students in the class will pass the exam is:

$$P(x \leq 1) = P(x = 0) + P(x = 1)$$

$$P(x \leq 1) = 0.0102 + 0.0768 = 0.0870$$

This looks like a good class!

Binomial Probability Tables

As the number of trials increases in a binomial experiment, calculating probabilities using the previous formula will really drain the batteries in your calculator and possibly even your brain. An easier way to arrive at these probabilities is to use a binomial probability table, which I have conveniently provided in Appendix B of this book. Following is an excerpt from this appendix, with the probabilities from our previous example underlined.

The probability table is organized by values of n, the total number of trials. The number of successes, x, are the rows of each section, whereas the probability of success, p, are the columns. Notice that the sum of each block of probabilities for a particular value of p adds to 1.0.

Values of p

n	x	0.1	0.2	0.3	0.4	0.5	0.6	0.7	0.8	0.9
4	0	0.6561	0.4096	0.2401	0.1296	0.0625	0.0256	0.0081	0.0016	0.0001
	1	0.2916	0.4096	0.4116	0.3456	0.2500	0.1536	0.0756	0.0256	0.0036
	2	0.0486	0.1536	0.2646	0.3456	0.3750	0.3456	0.2646	0.1536	0.0486
	3	0.0036	0.0256	0.0756	0.1536	0.2500	0.3456	0.4116	0.4096	0.2916
	4	0.0001	0.0016	0.0081	0.0256	0.0625	0.1296	0.2401	0.4096	0.6561
5	0	0.5905	0.3277	0.1681	0.0778	0.0313	0.0102	0.0024	0.0003	0.0000
	1	0.3280	0.4096	0.3601	0.2592	0.1563	0.0768	0.0284	0.0064	0.0005
	2	0.0729	0.2048	0.3087	0.3456	0.3125	0.2304	0.1323	0.0512	0.0081
	3	0.0081	0.0512	0.1323	0.2304	0.3125	0.3456	0.3087	0.2048	0.0729
	4	0.0005	0.0064	0.0283	0.0768	0.1563	0.2592	0.3601	0.4096	0.3281
	5	0.0000	0.0003	0.0024	0.0102	0.0313	0.0778	0.1681	0.3277	0.5905

One limitation of using binomial tables is that you are restricted to using only the values of p that are shown in the table. For instance, the previous table would not be useful for $p = 0.35$. Other statistics books may contain binomial tables that are more extensive than the one in Appendix B.

Using Excel to Calculate Binomial Probabilities

A convenient way to calculate binomial probabilities is to rely on our friend Excel, with its BINOM.DIST function. This built-in function has the following characteristics:

BINOM.DIST(x, n, p, cumulative)

where:

cumulative = FALSE if you want the probability of exactly x successes

cumulative = TRUE if you want the probability of x or fewer successes

For instance, Figure 9.2 shows the BINOM.DIST function being used to calculate the probability that two students out of the class of five will pass the exam.

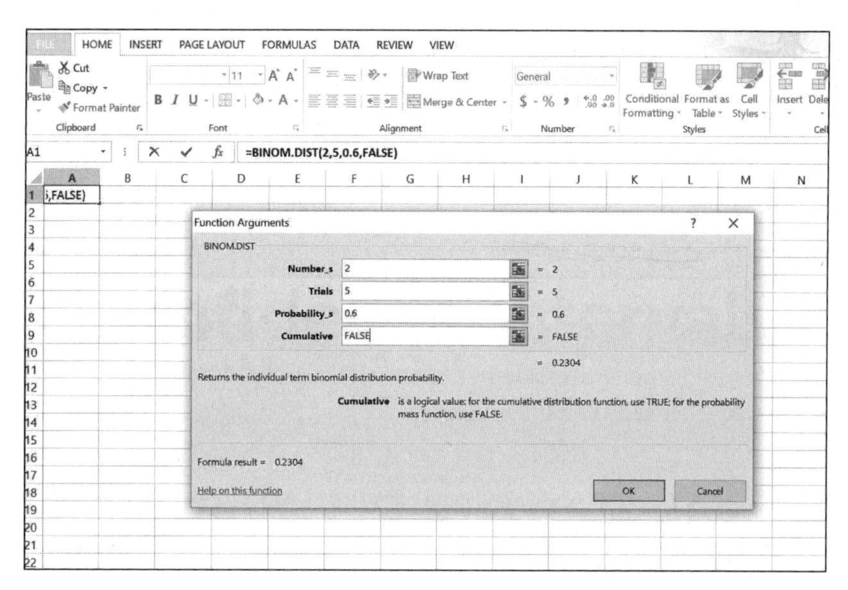

Figure 9.2
BINOM.DIST function in Excel for exactly x successes.

Cell A1 contains the Excel formula =BINOM.DIST(2,5,0.6,FALSE) with the result being 0.2304.

Excel will also calculate the probability that no more than two students out of the class of five will pass the exam, as shown in Figure 9.3.

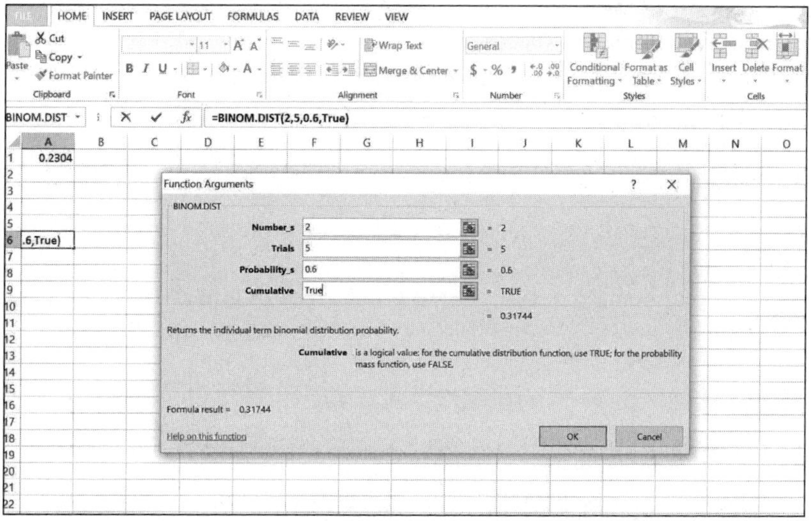

Figure 9.3
BINOM.DIST function in Excel for no more than x successes.

Cell A6 contains the Excel formula =BINOM.DIST(2,5,0.6,TRUE) with the result being 0.31744, which is the same as this:

$$P(x \leq 2) = P(x = 0) + P(x = 1) + P(x = 2)$$

$$P(x \leq 2) = 0.0102 + 0.0768 + 0.2304 = 0.3174$$

In other words, there is about 32 percent chance that no more than two students out of the class of 5 will pass the exam.

One benefit of using Excel to determine binomial probabilities is that you are not limited to the values of p shown in the binomial table in Appendix B. Excel's BINOM.DIST function allows you to use any value between 0 and 1 for p.

The Mean and Standard Deviation for the Binomial Distribution

You can calculate the mean for a binomial probability distribution by using the following equation:

$$\mu = np$$

where:

n = the number of trials

p = the probability of a success

For our example, the mean of the distribution is as follows:

$$\mu = np = (5)(0.6) = 3$$

In other words, on average, 3 out of the class of five students will pass the exam.

You can calculate the standard deviation for a binomial probability distribution using the following equation:

$$\sigma = \sqrt{npq}$$

where:

q = the probability of a failure

For our example, the standard deviation of the distribution is as follows:

$$\sigma = \sqrt{npq} = \sqrt{(5)(0.6)(0.4)} = 1.095$$

Well, that about covers the binomial probability distribution discussion. Don't be too sad, though; you'll see this again in future chapters.

Practice Problems

1. What is the probability of seeing exactly 7 heads after tossing a coin 10 times?
2. Goldey-Beacom College accepts 75 percent of applications that are submitted for entrance. What is the probability that they will accept exactly three of the next six applications?
3. Michael Jordan makes 80 percent of his free throws. What is the probability that he will make at least six of his next eight free-throw attempts?
4. A student randomly guesses at a 12-question, multiple-choice test where each question has 5 choices. What is the probability that the student will correctly answer exactly six questions?
5. Historical records show that 5 percent of people who visit a particular website purchase something. What is the probability that no more than two people out of the next seven will purchase something?
6. During the 2005 Major League Baseball season, Derrek Lee had a 0.335 batting average. Construct a binomial probability distribution for the number of successes (hits) for four official at bats during this season.
7. Sixty percent of a particular college population are female students. What is the probability that a class of 10 students has exactly 4 female students?

The Least You Need to Know

- A binomial experiment has only two possible outcomes for each trial.
- For a binomial experiment, the probability of success and failure is constant.
- Each trial of a binomial experiment is independent of any other trial in the experiment.

- The probability of x successes in n trials using the binomial distribution is as follows:

$$P(x) = \frac{n!}{x!(n-x)!} p^x q^{(n-x)}$$

- Calculate the mean for a binomial probability distribution by using the equation $\mu = np$.
- Calculate the standard deviation for a binomial probability distribution by using the equation $\sigma = \sqrt{npq}$.

CHAPTER

10

The Poisson Probability Distribution

Now that we have mastered the binomial probability distribution, we are ready to move on to the next discrete theoretical distribution, the Poisson. This probability distribution is named after Simeon Poisson, a French mathematician who developed the distribution during the early 1800s.

The Poisson distribution is useful for calculating the probability that a certain number of events will occur over a specific period of time. We could use this distribution to determine the likelihood that 10 customers will walk into a store during the next hour or that 2 car accidents will occur at a busy intersection this month. So let's grab some crêpes and croissants and learn about some French math.

In This Chapter

- Describe the characteristics of a Poisson process
- Calculate probabilities using the Poisson equation
- Use the Poisson probability tables
- Use Excel to calculate Poisson probabilities
- Use the Poisson equation to approximate the binomial equation

Characteristics of a Poisson Process

In Chapter 9, we defined a binomial experiment as counting the number of successes over a specific number of trials. The result of each trial is either a success or a failure. A *Poisson process* counts the number of occurrences of an event over a period of time, area, distance, or any other type of measurement.

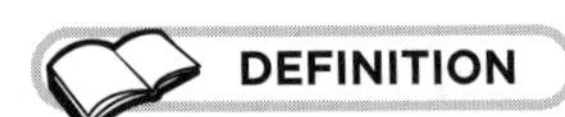

A **poisson process** has the following characteristics: (1) the experiment consists of counting the number of occurrences of an event over a period of time, area, distance, or any other type of measurement; (2) the mean of the Poisson distribution has to be the same for each interval of measurement; (3) the number of occurrences during one interval is independent of the number of occurrences in any other interval; (4) the intervals don't overlap.

Rather than being limited to only two outcomes, the Poisson process can have any number of outcomes over the unit of measurement. For instance, the number of customers who walk into a local convenience store during the next hour could be zero, one, two, three, or so on. The random variable for the Poisson distribution would be the actual number of occurrences—in this case, the number of customers arriving during the next hour.

The mean for a Poisson distribution is the average number of occurrences that would be expected over the unit of measurement. For a Poisson process, the mean has to be the same for each interval of measurement. For instance, if the average number of customers walking into the store each hour is 11, this average needs to apply to every one-hour increment.

Another characteristic of a Poisson process is that the number of occurrences during one interval is independent of the number of occurrences in other intervals. In other words, if six customers walk into the store during the first hour of business, this would have no effect on the number of customers arriving during the second hour.

The last characteristic of a Poisson process is that the intervals don't overlap. For example, when counting the number of customers walking into the store in one hour periods, the one-hour periods cannot overlap with each other. We can count the number of customers arriving between 9 to 10 A.M. and 10 to 11 A.M. and so forth, but we cannot use the 9:30 to 10:30 A.M. period because this overlaps with the other intervals.

Examples of random variables that may follow a Poisson probability distribution include the following:

- The number of cars that arrive at a tollbooth over a specific period of time
- The number of typographical errors found in a manuscript

- The number of students who are absent in my Monday morning statistics class
- The number of professional football players who are placed on the injured list each week

Now that you understand the basics of a Poisson process, let's move into probability calculations.

The Poisson Probability Distribution

If a random variable follows a pattern consistent with a Poisson probability distribution, then we can calculate the probability of a certain number of occurrences over a given interval. To make this calculation, we need to know the average number of occurrences for the event over this interval. To demonstrate the use of the Poisson probability distribution, I'll use this example.

The following story is true, but the names have not been changed because nobody in this story is innocent. Each year, Brian, John, and Bob make a golf pilgrimage to Myrtle Beach, South Carolina. On their last night one particular year, they were browsing through a golf store. Brian somehow convinced Bob to purchase a used, fancy, brand-name golf club that he swore he absolutely had to have in order to reach his full potential as a golfer. Even used, this club cost more than any Bob had purchased new, but teenagers have this special talent that allows them to disregard any rational adult logic when their minds are made up.

Early the following morning, they packed their bags, checked out of the hotel, and drove to their final round of golf, which Bob had cleverly planned to be along their route back home. On the first tee, Brian pulled out his new, used prize possession and proceeded to hit a "duck hook," which is a golfer's term for a ball that goes very short and very left, often into a bunch of trees never to be seen again. Bob smiled nervously at Brian and tried to convince himself that he'd be fine on the next hole. After hitting duck hooks on holes two, three, and four, Bob found himself physically restraining Brian from throwing his new, used prized possession into the lake.

After their round was over, Bob drove *back* to Myrtle Beach to return the club, adding an hour to what would have been a 10-hour car ride. At the golf store, the woman cheerfully said she would take the club back, but she needs … *the receipt.* Now Bob vaguely remembered putting the receipt someplace "special" just in case he would need it, but after packing, checking out, and playing golf, he would have had a better chance of discovering a cure for cancer than remembering where he had put that piece of paper.

Not being one to give up easily, Bob marched back to the car and started unpacking everything. After a short while, the same woman walked out to tell Bob the store would gladly refund his money *without the receipt* if he would just pack up his things and put them back in the car. A very powerful technique that Bob discovered when he needed to return something without a receipt, simply take along some dirty clothes in a suitcase and spread them out in the parking lot in front of the store. It works like a charm.

Anyway, let's assume that the number of tee shots that Brian normally hits that actually land in the fairway during a round of golf is five. The fairway is the area of short grass where the people who have designed this nerve-wracking game intended your tee shot to land. We will also assume that the actual number of fairways that Brian "hits" during one round follows the Poisson distribution.

WRONG NUMBER

How do I know that the actual number of fairways that Brian "hits" during one round follows the Poisson distribution? At this point, I really don't know for sure. What I would need to do to verify this claim is record the number of fairways hit over several rounds and then perform a "Goodness of Fit" test to decide whether the data fits the pattern of a Poisson distribution. I promise you that we will perform this test in Chapter 18, so please be patient.

We can now use the Poisson probability distribution to calculate the probability that Brian will hit x number of fairways during his next round, as follows:

$$P(x) = \frac{\lambda^x e^{-\lambda}}{x!}$$

where:

x = the number of occurrences of interest over the interval

λ = the mean number of occurrences over the interval

e = the mathematical constant ≈ 2.71828

$P(x)$ = the probability of exactly x occurrences over the interval

We can now calculate the probability that Brian will hit exactly seven fairways during his next round. With $\lambda = 5$, the equation becomes this:

$$P(x=7) = \frac{(5)^7 (2.71828)^{-5}}{7!}$$

$$P(x=7) = \frac{(78125)(0.006738)}{7 \cdot 6 \cdot 5 \cdot 4 \cdot 3 \cdot 2 \cdot 1} = 0.1044$$

In other words, Brian has slightly more than a 10 percent chance of hitting exactly seven fairways.

We can also calculate the cumulative probability that Brian will hit no more than two fairways using the following equations:

$$P(x \le 2) = P(x = 0) + P(x = 1) + P(x = 2)$$

$$P(x=0) = \frac{(5)^0(2.71828)^{-5}}{0!} = \frac{(1)(0.006738)}{1} = 0.0067$$

$$P(x=1) = \frac{(5)^1(2.71828)^{-5}}{1!} = \frac{(5)(0.006738)}{1} = 0.0337$$

$$P(x=2) = \frac{(5)^2(2.71828)^{-5}}{2!} = \frac{(25)(0.006738)}{2 \cdot 1} = 0.0842$$

$$P(x \le 2) = 0.0067 + 0.0337 + 0.0842 = 0.1246$$

There is a 12.46 percent chance that Brian will hit no more than two fairways during his next round.

In the previous example, the mean of the Poisson distribution happened to be an integer (5). However, this doesn't have to always be the case. Suppose the number of absent students for my Monday morning statistics follows a Poisson distribution, with the average being 2.4 students. The probability that there will be three students absent next Monday is as follows.

$$P(x=3) = \frac{(2.4)^3(2.71828)^{-2.4}}{3!}$$

$$P(x=3) = \frac{(13.824)(0.090718)}{3 \cdot 2 \cdot 1} = 0.2090$$

Looks like I need to start taking roll on Mondays!

There's one more cool feature of the Poisson distribution: the variance of the distribution is the same as the mean. In other words:

$$\sigma^2 = \lambda$$

This means that there are no nasty variance calculations like the ones we dealt with in previous chapters for this distribution.

Poisson Probability Tables

Just like the binomial distribution, the Poisson probability distribution has a table that allows you to look up the probabilities for certain mean values. You can find the Poisson distribution table in Appendix B of this book. The following is an excerpt from this appendix with the probabilities from our Myrtle Beach example underlined.

Values of λ

x	3.2	3.4	3.6	3.8	4.0	4.2	4.4	4.6	4.8	5.0
0	0.0408	0.0334	0.0273	0.0224	0.0183	0.0150	0.0123	0.0101	0.0082	0.0067
1	0.1304	0.1135	0.0984	0.0850	0.0733	0.0630	0.0540	0.0462	0.0395	0.0337
2	0.2087	0.1929	0.1771	0.1615	0.1465	0.1323	0.1188	0.1063	0.0948	0.0842
3	0.2226	0.2186	0.2125	0.2046	0.1954	0.1852	0.1743	0.1631	0.1517	0.1404
4	0.1781	0.1858	0.1912	0.1944	0.1954	0.1944	0.1917	0.1875	0.1820	0.1755
5	0.1140	0.1264	0.1377	0.1477	0.1563	0.1633	0.1687	0.1725	0.1747	0.1755
6	0.0608	0.0716	0.0826	0.0936	0.1042	0.1143	0.1237	0.1323	0.1398	0.1462
7	0.0278	0.0348	0.0425	0.0508	0.0595	0.0686	0.0778	0.0869	0.0959	0.1044
8	0.0111	0.0148	0.0191	0.0241	0.0298	0.0360	0.0428	0.0500	0.0575	0.0653
9	0.0040	0.0056	0.0076	0.0102	0.0132	0.0168	0.0209	0.0255	0.0307	0.0363
10	0.0013	0.0019	0.0028	0.0039	0.0053	0.0071	0.0092	0.0118	0.0147	0.0181
11	0.0004	0.0006	0.0009	0.0013	0.0019	0.0027	0.0037	0.0049	0.0064	0.0082
12	0.0001	0.0002	0.0003	0.0004	0.0006	0.0009	0.0013	0.0019	0.0026	0.0034
13	0.0000	0.0000	0.0001	0.0001	0.0002	0.0003	0.0005	0.0007	0.0009	0.0013
14	0.0000	0.0000	0.0000	0.0000	0.0001	0.0001	0.0001	0.0002	0.0003	0.0005
15	0.0000	0.0000	0.0000	0.0000	0.0000	0.0000	0.0000	0.0001	0.0001	0.0002

The probability table is organized by values of λ, the average number of occurrences. Notice that the sum of each block of probabilities for a particular value of λ adds to 1.

As with the binomial tables, one limitation of using the Poisson tables is that you are restricted to using only the values of λ that are shown in the table. For instance, the previous table would not be useful for $\lambda = 4.5$. However, other statistics books might contain Poisson tables that are more extensive than the one in Appendix B.

The Poisson distribution for $\lambda = 5$ is shown graphically in the following histogram. The probabilities in Figure 10.1 are taken from the last column in the previous table.

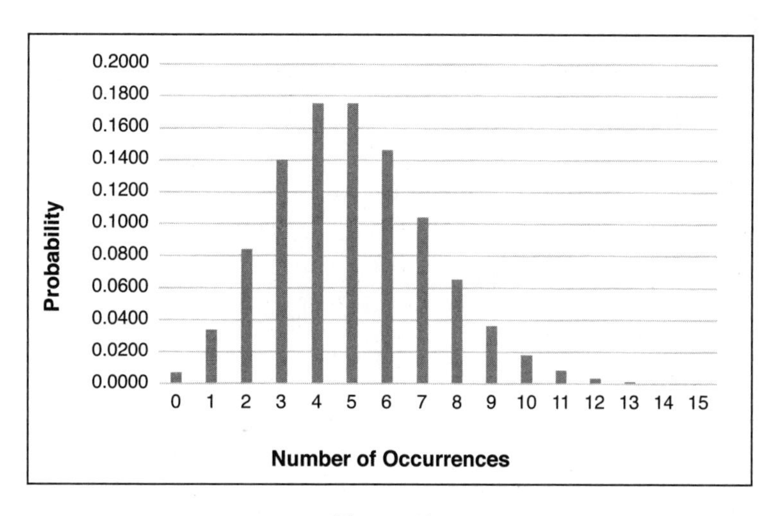

Figure 10.1
Poisson probability distribution.

Note that the most likely numbers of occurrences for this distribution are four and five.

Here's another example. Let's assume that the number of car accidents each month at a busy intersection that Bob used to pass on his way to work follows the Poisson distribution with a mean of 1.8 accidents per month. What is the probability that three or more accidents will occur next month? You can express this as:

$$P(x \geq 3) = P(x = 3) + P(x = 4) + P(x = 5) + \quad + P(x = \infty)$$

Technically, with a Poisson distribution, there is no upper limit to the number of occurrences during the interval. You'll notice from the Poisson tables that the probability of a large number of occurrences is practically zero. Because we cannot add all the probabilities of an infinite number of occurrences, we need to use the complement rule (Do you remember this from Chapter 6? It all comes back!), which is:

$$P(x \geq 3) = 1 - P(x < 3)$$

because:

$$P(x = 0) + P(x = 1) + P(x = 2) + P(x = 3) + \quad + P(x = \infty) = 1.0$$

Therefore, to find the probability of three or more accidents, we'll use the following:

$$P(x \geq 3) = 1 - [P(x = 0) + P(x = 1) + P(x = 2)]$$

Using the probabilities underlined in the following Poisson table (I seem to have misplaced my calculator), we have this:

Values of λ

x	1.1	1.2	1.3	1.4	1.5	1.6	1.7	1.8	1.9	2.0
0	0.3329	0.3012	0.2725	0.2466	0.2231	0.2019	0.1827	<u>0.1653</u>	0.1496	0.1353
1	0.3662	0.3614	0.3543	0.3452	0.3347	0.3230	0.3106	<u>0.2975</u>	0.2842	0.2707
2	0.2014	0.2169	0.2303	0.2417	0.2510	0.2584	0.2640	<u>0.2678</u>	0.2700	0.2707
3	0.0738	0.0867	0.0998	0.1128	0.1255	0.1378	0.1496	0.1607	0.1710	0.1804
4	0.0203	0.0260	0.0324	0.0395	0.0471	0.0551	0.0636	0.0723	0.0812	0.0902
5	0.0045	0.0062	0.0084	0.0111	0.0141	0.0176	0.0216	0.0260	0.0309	0.0361
6	0.0008	0.0012	0.0018	0.0026	0.0035	0.0047	0.0061	0.0078	0.0098	0.0120
7	0.0001	0.0002	0.0003	0.0005	0.0008	0.0011	0.0015	0.0020	0.0027	0.0034
8	0.0000	0.0000	0.0001	0.0001	0.0001	0.0002	0.0003	0.0005	0.0006	0.0009
9	0.0000	0.0000	0.0000	0.0000	0.0000	0.0000	0.0001	0.0001	0.0001	0.0002

$$P(x \geq 3) = 1 - (0.1653 + 0.2975 + 0.2678)$$

$$P(x \geq 3) = 1 - 0.7306 = 0.2694$$

There is almost a 27 percent chance that three or more accidents will occur in this intersection next month.

Using Excel to Calculate Poisson Probabilities

You can also conveniently calculate Poisson probabilities using Excel. The built-in POISSON.DIST function has the following characteristics:

POISSON.DIST(x, λ, cumulative)

where:

cumulative = FALSE if you want the probability of exactly x occurrences

cumulative = TRUE if you want the probability of x or fewer occurrences

For instance, Figure 10.2 shows the POISSON.DIST function being used to calculate the probability that there will be exactly two accidents in the intersection next month.

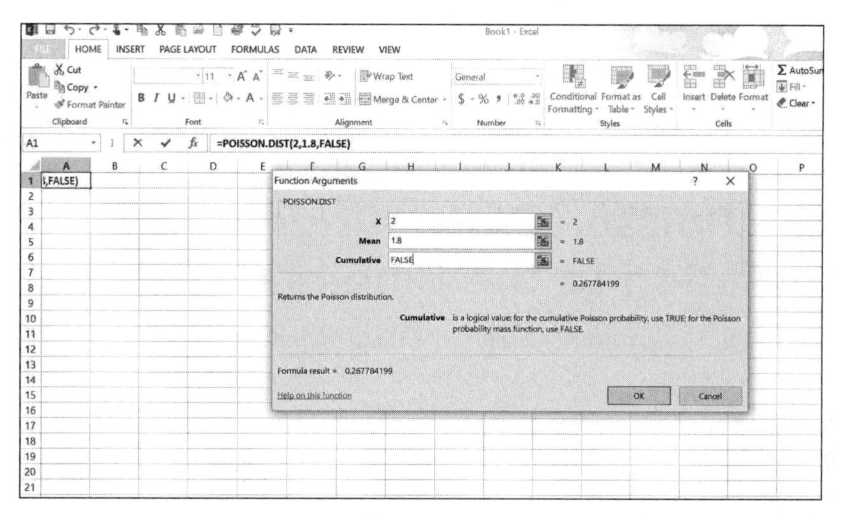

Figure 10.2

POISSON.DIST function in Excel for exactly x occurrences.

Cell A1 contains the Excel formula =POISSON.DIST(2,1.8,FALSE) with the result being 0.2678. This probability is underlined in the previous table.

Excel will also calculate the cumulative probability that there will be no more than two accidents in the intersection, as shown in Figure 10.3.

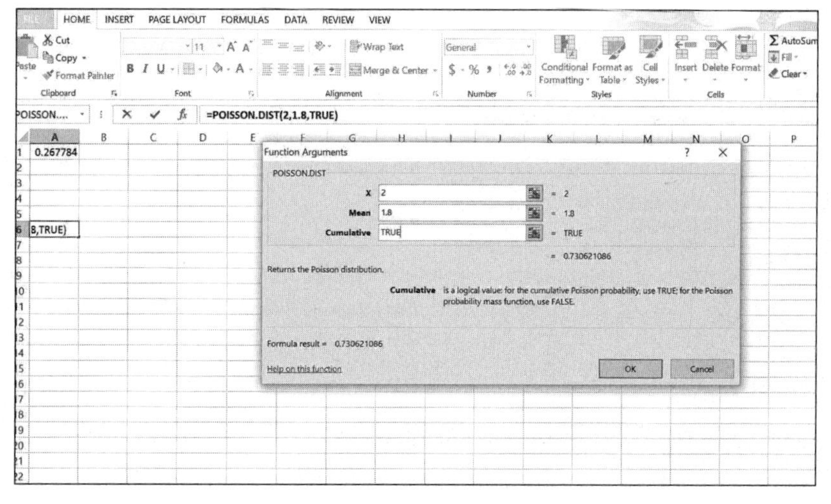

Figure 10.3

POISSON.DIST function in Excel for no more than x occurrences.

Cell A6 contains the Excel formula =POISSON.DIST(2,1.8,TRUE) with the result being 0.7306, a probability that we saw in the last calculation and which is also the sum of the underlined probabilities in the previous table.

One benefit of using Excel to determine Poisson probabilities is that you are not limited to the values of λ shown in the Poisson table in Appendix B. Excel's POISSON.DIST function allows you to use any value for λ.

Using the Poisson Distribution as an Approximation to the Binomial Distribution

I don't know about you, but when I have two ways to do something, I like to choose the one that's less work. If you don't agree with me, feel free to skip this material. If you do, read on!

We can use the Poisson distribution to calculate binomial probabilities under the following conditions:

- when the number of trials, n, is greater than or equal to 20, and
- when the probability of a success, p, is less than or equal to 0.05.

The Poisson formula would look like this:

$$P(x) = \frac{(np)^x e^{-(np)}}{x!}$$

where:

n = the number of trials

x = the number of successes

p = the probability of a success

BOB'S BASICS

If you need to calculate binomial probabilities with the number of trials, n, greater than or equal to 20 and the probability of a success, p, less than or equal to 0.05, then you can use the equation for the Poisson distribution to approximate the binomial probabilities.

You might be asking yourself at this moment just why you would want to do this. The answer is because the Poisson formula has fewer computations than the binomial formula, and, under the stated conditions, the distributions are very close to one another.

Just in case you are from Missouri (the "Show Me" state), I'll demonstrate this point with an example. Suppose there are 20 traffic lights in a town and each has a 3 percent chance of not working properly (a success) on any given day. What is the probability that exactly 1 of the 20 lights will not work today? This is a binomial experiment with $n = 20$, $x = 1$, and $p = 0.03$. From Chapter 9, we know that the binomial probability is this:

$$P(x) = \frac{n!}{x!(n-x)!} p^x q^{(n-x)}$$

$$P(1) = \frac{20!}{1!(20-1)!} (0.03)^1 (0.97)^{20-1}$$

$$P(1) = (20)(0.03)(0.560613) = 0.3364$$

The Poisson approximation is as follows:

$$P(x) = \frac{(np)^x e^{-(np)}}{x!}$$

Because: np = (20)(0.03) = 0.6

$$P(1) = \frac{(0.6)^1 e^{-0.6}}{1!}$$

$$P(1) = (0.6)(0.548812) = 0.3293$$

Even if you're from Missouri, I think you would have to agree that the Poisson calculation is easier, and the two results are very close. So my advice to you is to use the Poisson equation if you're faced with calculating binomial probabilities with $n \geq 20$ and $p \leq 0.05$.

This concludes our discussion of discrete probability distributions. We hope you've had as much fun with these as we've had!

Practice Problems

1. The number of rainy days per month at a particular town follows a Poisson distribution with a mean value of six days. What is the probability that it will rain four days next month?
2. The number of customers arriving at a particular store follows a Poisson distribution with a mean value of 7.5 customers per hour. What is the probability that five customers will arrive during the next hour?
3. The number of pieces of mail that I receive daily follows a Poisson distribution with a mean value of 4.2 per day. What is the probability that I will receive more than two pieces of mail tomorrow?

4. The number of employees who call in sick on Monday follows a Poisson distribution with a mean value of 3.6. What is the probability that no more than three employees will call in sick next Monday?

5. The number of spam e-mails that I receive each day follows a Poisson distribution with a mean value of 2.5. What is the probability that I will receive exactly one spam e-mail tomorrow?

6. Historical records show that 5 percent of people who visit a particular website purchase something. What is the probability that exactly 2 people out of the next 25 will purchase something? Use the Poisson distribution to estimate this binomial probability.

The Least You Need to Know

- A Poisson process counts the number of occurrences of an event over a period of time, area, distance, or any other type of measurement.
- The mean for a Poisson distribution is the average number of occurrences that would be expected over the unit of measurement and has to be the same for each interval of measurement.
- The number of occurrences during one interval of a Poisson process is independent of the number of occurrences in other intervals.
- The intervals of a Poisson process don't overlap.
- If x is a Poisson random variable, the probability of x occurrences over the interval of measurement is $P(x) = \frac{\lambda^x e^{-\lambda}}{x!}$.
- If the number of binomial trials is greater than or equal to 20 and the probability of a success is less than or equal to 0.05, then you can use the equation for the Poisson distribution to approximate the binomial probabilities.

CHAPTER

11

The Normal Probability Distribution

Now let's take on a new challenge: continuous random variables and a continuous probability distribution known as the normal distribution. Remember that in Chapter 8 we defined a continuous random variable as one that can assume any numerical value within an interval as a result of measuring the outcome of an experiment. Some examples of continuous random variables are weight, distance, speed, and time.

The normal distribution is a statistician's workhorse. This distribution is the foundation for many types of inferential statistics that we rely on today. We will continue to refer to this distribution through many of the remaining chapters in this book.

In This Chapter

- Examining the properties of a normal probability distribution
- Using the standard normal table to calculate probabilities of a normal random variable
- Using Excel to calculate normal probabilities
- Using the normal distribution as an approximation to the binomial distribution

Characteristics of the Normal Probability Distribution

A continuous random variable that follows the normal probability distribution has several distinctive features. Let's say that the monthly rainfall in inches for a particular city follows the normal distribution with an average of 3.5 inches and a standard deviation of 0.8 inches. The probability distribution for such a random variable is shown in Figure 11.1.

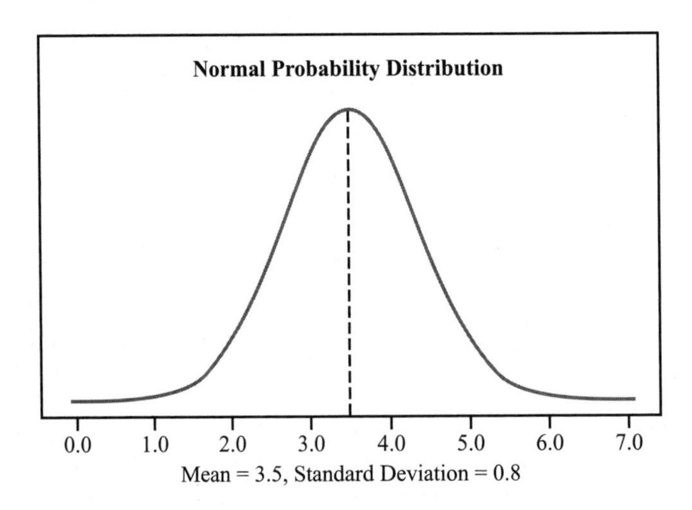

Figure 11.1
Normal probability distribution with a mean = 3.5, a standard deviation = 0.8.

From this figure, we can make the following observations about the normal distribution:

- The distribution is bell-shaped and symmetrical around the mean.
- The mean, median, and mode are the same value—3.5 inches—and in this case fall in the exact center of the distribution.
- The total area under the curve is equal to 1; half of the area is to the right of the mean and half of it is to the left.
- The distribution is *asymptotic*, meaning that the left and right tails of the normal probability distribution extend indefinitely, never quite touching the horizontal axis.

The mean and the standard deviation determine the position and the shape of the normal distribution. The standard deviation plays an important role in the shape of the curve. Looking at Figure 11.1, we can see that nearly all of the monthly rainfall measurements would fall between 1.0 and 6.0 inches. Now look at Figure 11.2, which shows the normal distribution with the same mean of 3.5 inches, but with a standard deviation of only 0.5 inches.

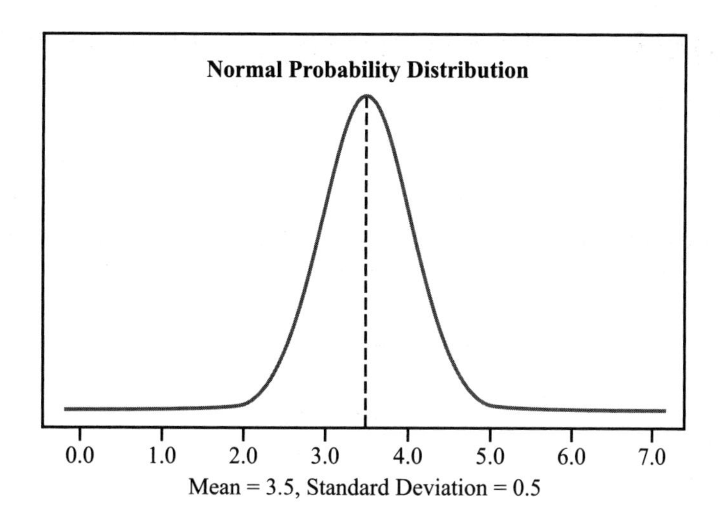

Figure 11.2
Normal probability distribution with a mean = 3.5, a standard deviation = 0.5.

Here you see a curve that's much tighter around the mean. Almost all the rainfall measurements will be between 2.0 and 5.0 inches per month.

The mean of the normal distribution determines the central point of the distribution. As the mean changes, so does the center point of the distribution. Figure 11.3 shows the impact of changing the mean of the distribution to 5.0 inches, leaving the standard deviation at 0.8 inches.

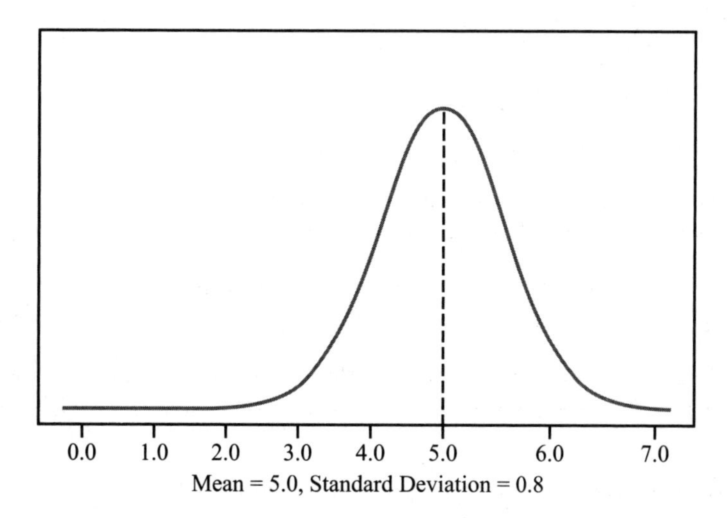

Figure 11.3
Normal probability distribution with a mean = 5.0, a standard deviation = 0.8.

BOB'S BASICS

A smaller standard deviation results in a "skinnier" curve that's tighter and taller around the mean. A larger standard deviation makes for a "fatter" curve that's more spread out and not as tall.

In each of the previous figures, the characteristics of the normal probability distribution hold true. In each case, the values of μ—the mean—and σ—the standard deviation—completely describe the position and shape of the distribution.

The probability function for the normal distribution has a particularly mean personality (that pun was surely intended) and is shown as follows:

$$f(x) = \frac{1}{\sigma\sqrt{2\pi}} e^{-(1/2)[(x- \quad)/\sigma]^2}$$

I promise you this will be the last you'll see of this beast. Fortunately, we have other methods for calculating probabilities for this distribution that are more civilized and which we will discuss in the next section.

Calculating Probabilities for the Normal Distribution

There are a couple of approaches to calculate probabilities for a normal random variable: tables and Excel. Let's start with the table and then move to Excel after that.

To help us calculate the probability for a normal random variable, statisticians introduced a table that we can use instead of the formula above. The problem that they encountered in creating a table is that each normal distribution has its own mean and standard deviation, so they could not come up with just one table to use. They would have to have many tables, each with its own mean and standard deviation. As you might imagine, this would be very tedious. Instead, statisticians said that we can standardize the normal distribution. This way, we can have just one standard normal table with the same mean and standard deviation. The standard normal distribution has a mean (μ) of 0 and standard deviation (σ) of 1. How did they do that? By using the z-score. To demonstrate, let's use the following example.

One morning a few years ago, Debbie called Bob on his cell phone while he was out running errands and spoke the two words that he had feared hearing. "They're back," she said. "Okay," he replied somberly, and then hung up the phone and headed straight toward the hardware store. His manhood was being challenged, and he'd be darned if he was going to take this lying down. This was *war*, and he was going home fully prepared for battle! He was referring to his annual struggle with the most vile, the most dastardly, the most hungry creature that God has ever placed on this planet … the Japanese beetle.

By the time Bob returned home from the hardware store, half of his beautiful plum tree looked like Swiss cheese. He quickly counterattacked with a vengeance, spraying the most potent chemicals money could buy. In the end, after the toxic spray cleared, he stood alone, master of his domain.

Alright, let's say that the amount of toxic spray Bob uses each year follows a normal distribution with a mean of 60 ounces and a standard deviation of 5 ounces. This means that each year Bob battles with these demons, the most likely amount of spray he uses is 60 ounces, but it will vary year to year. The probability of other amounts above and below 60 ounces will drop off according to the bell-shaped curve. Armed with this information, we are now ready to determine probabilities of various usages each year.

Calculating Probability Using the Z-Score

Because the total area under a normal distribution curve equals 1 and the curve is symmetrical, we can say the probability that Bob uses 60 ounces or more of spray is 50 percent, as is the probability that he uses 60 ounces or less. This is shown in Figure 11.4.

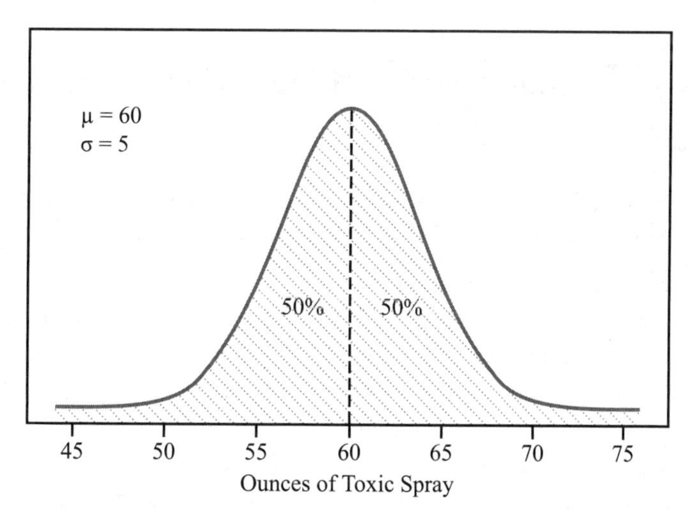

Figure 11.4
Normal probability distribution with a mean = 60, a standard deviation = 5.0.

How would you calculate the probability that Bob would use 64.3 ounces of spray or less the following year? I'm glad you asked. For this task, we need to use the *standard normal* distribution we mentioned above, which is a normal distribution with $\mu = 0$ and $\sigma = 1$, and is shown in Figure 11.5.

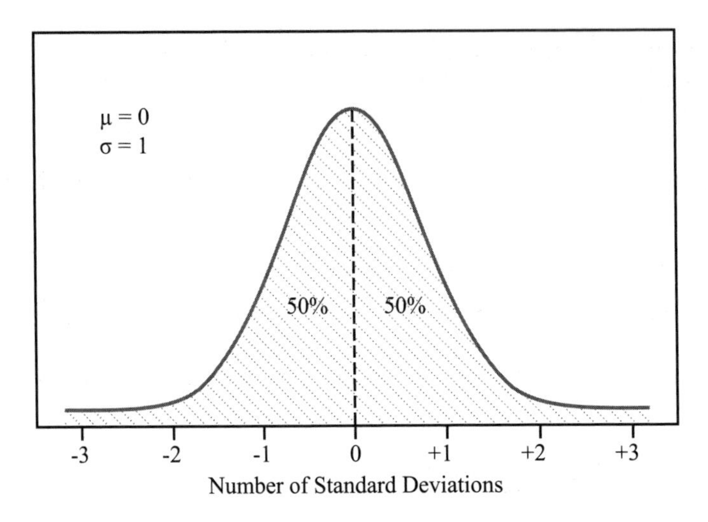

Figure 11.5
Standard normal probability distribution with a mean = 0, a standard deviation = 1.0.

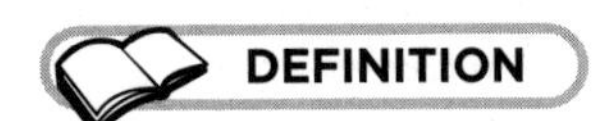

The **standard normal** distribution is a standardized distribution with a mean equal to 0 and a standard deviation equal to 1.0.

This standard normal distribution is the basis for all normal probability calculations, and we'll use it throughout this chapter. To standardize the normal distribution, we need to calculate the z-score. What is the z-score? Read on!

The z-score is the number of standard deviations any point is away from the mean. So, our next step is to determine how many standard deviations the value 64.3 is from the mean of 60 and show this value on the standard normal distribution curve. We do this using the following formula:

$$z = \frac{x-\mu}{\sigma}$$

where:

x = the normally distributed random variable of interest

μ = the mean of the normal distribution

σ = the standard deviation of the normal distribution

z = the number of standard deviations between x and μ, otherwise known as the standard *z-score*

For our example, the standard z-score is as follows:

$$z = \frac{64.3 - 60}{5} = 0.86$$

Now I know that 64.3 is 0.86 standard deviations away from 60 in my distribution.

TEST YOUR KNOWLEDGE

Can the z-score be negative? Yes! The z-score is negative for all x values to the left of the mean (less than the mean), positive for all x values to the right of the mean, and the z-score is 0 when the x value is the same as the mean.

Using the Standard Normal Table

Now that we have the standard z-score, we can use the following table to determine the probability that Bob uses 64.3 ounces of toxic spray or less the next year. This table is an excerpt from Appendix B and shows the area of the standard normal curve up to and including certain values of z. Because $z = 0.86$ in our example, we go to the 0.8 row and the 0.06 column to find the value 0.8051, which is underlined. This is the probability. Yes, it's that easy!

Second digit of z

z	0.00	0.01	0.02	0.03	0.04	0.05	0.06	0.07	0.08	0.09
0.0	0.5000	0.5040	0.5080	0.5120	0.5160	0.5199	0.5239	0.5279	0.5319	0.5359
0.1	0.5398	0.5438	0.5478	0.5517	0.5557	0.5596	0.5636	0.5675	0.5714	0.5753
0.2	0.5793	0.5832	0.5871	0.5910	0.5948	0.5987	0.6026	0.6064	0.6103	0.6141
0.3	0.6179	0.6217	0.6255	0.6293	0.6331	0.6368	0.6406	0.6443	0.6480	0.6517
0.4	0.6554	0.6591	0.6628	0.6664	0.6700	0.6736	0.6772	0.6808	0.6844	0.6879
0.5	0.6915	0.6950	0.6985	0.7019	0.7054	0.7088	0.7123	0.7157	0.7190	0.7224
0.6	0.7257	0.7291	0.7324	0.7357	0.7389	0.7422	0.7454	0.7486	0.7517	0.7549
0.7	0.7580	0.7611	0.7642	0.7673	0.7704	0.7734	0.7764	0.7794	0.7823	0.7852
0.8	0.7881	0.7910	0.7939	0.7967	0.7995	0.8023	<u>0.8051</u>	0.8078	0.8106	0.8133
0.9	0.8159	0.8186	0.8212	0.8238	0.8264	0.8289	0.8315	0.8340	0.8365	0.8389
1.0	0.8413	0.8438	0.8461	0.8485	0.8508	0.8531	0.8554	0.8577	0.8599	0.8621

This area is shown graphically in Figure 11.6.

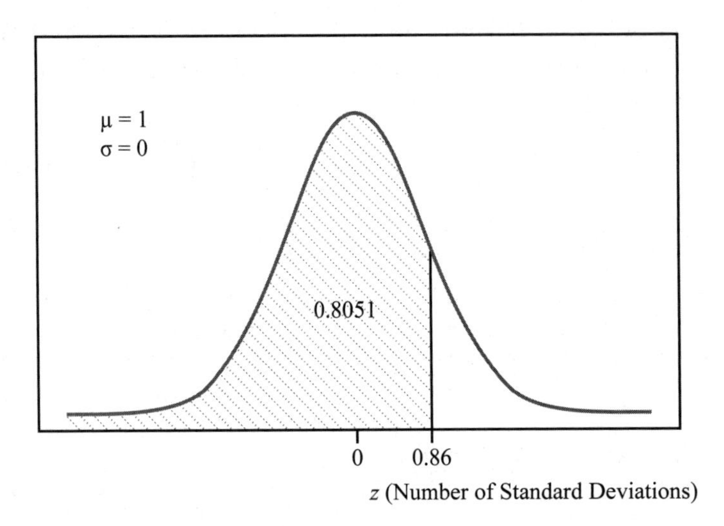

Figure 11.6
The shaded area represents the probability that z will be less than or equal to 0.86.

The probability that the standard *z*-score will be less than or equal to 0.86 is 80.51 percent. Because:

$$P(x \leq 64.3) = P(z \leq 0.86) = 0.8051$$

There is an 80.51 percent chance Bob uses 64.3 ounces of spray or less the next year against those evil Japanese beetles. This can be seen in Figure 11.7.

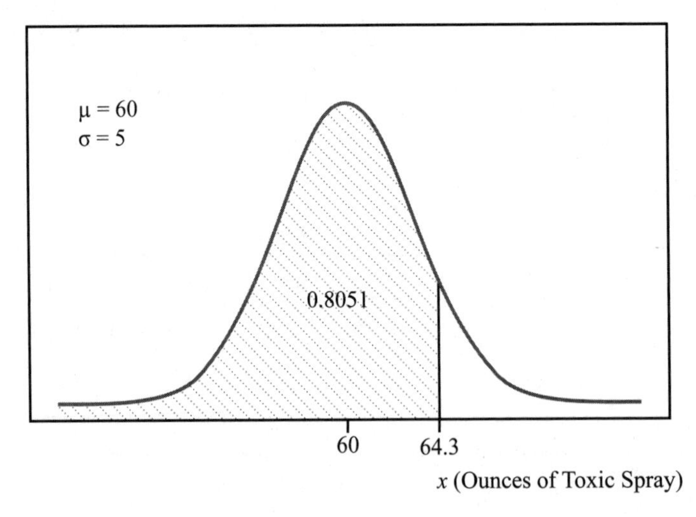

Figure 11.7
The shaded area represents the probability that x will be less than or equal to 64.3 ounces.

What about the probability that Bob uses more than 62.5 ounces of spray the next year? Because the standard normal table only has probabilities that are less than or equal to the *z*-scores, we need to look at the complement to this event.

$$P(x > 62.5) = 1 - P(x \leq 62.5)$$

The *z*-score now becomes this:

$$z = \frac{62.5-60}{5} = 0.50$$

According to the standard normal table:

$$P(z \leq 0.50) = 0.6915$$

But we want:

$$P(z > 0.50) = 1 - 0.6915 = 0.3085$$

This probability is shown graphically in Figure 11.8.

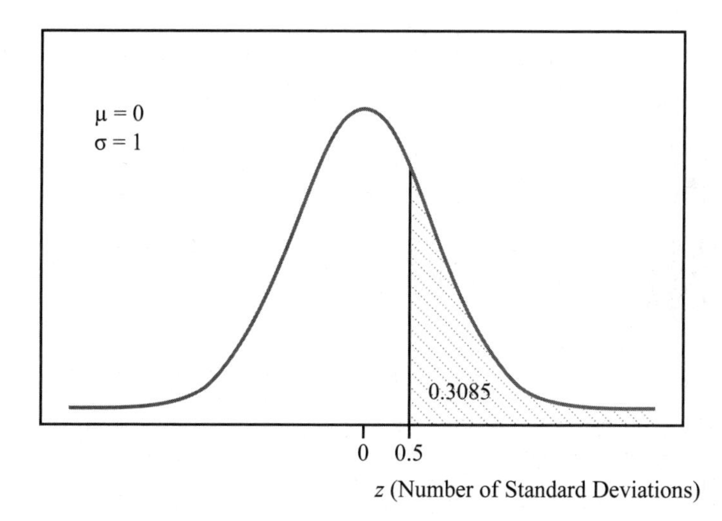

Figure 11.8
The shaded area represents the probability that z will be more than 0.50 ounces.

Because:

$$P(x > 62.5) = P(z > 0.50) = 0.3085$$

There is a 30.85 percent chance that Bob uses more than 62.5 ounces of toxic spray. Beetles beware!

What about the probability that Bob uses more than 54 ounces of spray? Again, we need the complement rule, which would be this:

$$P(x > 54) = 1 - P(x \leq 54)$$

The z-score becomes this:

$$z = \frac{54-60}{5} = -1.20$$

The negative z-score indicates that we are to the left of the distribution mean. Now we are going to look at the other standard normal distribution table, which has the negative values of z. The table below is an excerpt from Appendix B and shows the negative values of z. In our example, z = -1.20, so we go to the -1.2 row and the 0.00 column to find the value 0.1151, which is underlined.

Second digit of z

z	0.00	0.01	0.02	0.03	0.04	0.05	0.06	0.07	0.08	0.09
-1.3	0.0968	0.0951	0.0934	0.0918	0.0901	0.0885	0.0869	0.0853	0.0838	0.0823
-1.2	<u>0.1151</u>	0.1131	0.1112	0.1093	0.1075	0.1056	0.1038	0.1020	0.1003	0.0985
-1.1	0.1357	0.1335	0.1314	0.1292	0.1271	0.1251	0.1230	0.1210	0.1190	0.1170
-1.0	0.1587	0.1562	0.1539	0.1515	0.1492	0.1469	0.1446	0.1423	0.1401	0.1379
-0.9	0.1841	0.1814	0.1788	0.1762	0.1736	0.1711	0.1685	0.1660	0.1635	0.1611
-0.8	0.2119	0.2090	0.2061	0.2033	0.2005	0.1977	0.1949	0.1922	0.1894	0.1867
-0.7	0.2420	0.2389	0.2358	0.2327	0.2296	0.2266	0.2236	0.2206	0.2177	0.2148
-0.6	0.2743	0.2709	0.2676	0.2643	0.2611	0.2578	0.2546	0.2514	0.2483	0.2451
-0.5	0.3085	0.3050	0.3015	0.2981	0.2946	0.2912	0.2877	0.2843	0.2810	0.2776
-0.4	0.3446	0.3409	0.3372	0.3336	0.3300	0.3264	0.3228	0.3192	0.3156	0.3121

This is the probability that x ≤ 54. What we need is the probability that x > 54, but we know what to do now! Use the complement rule as follows:

$$P(x > 54) = P(z > -1.2) = 1 - P(z \leq -1.2) = 1 - 0.1151 = 0.8849$$

There is an 88.49 percent chance Bob will use more than 54 ounces of spray. This probability is shown graphically in Figure 11.9.

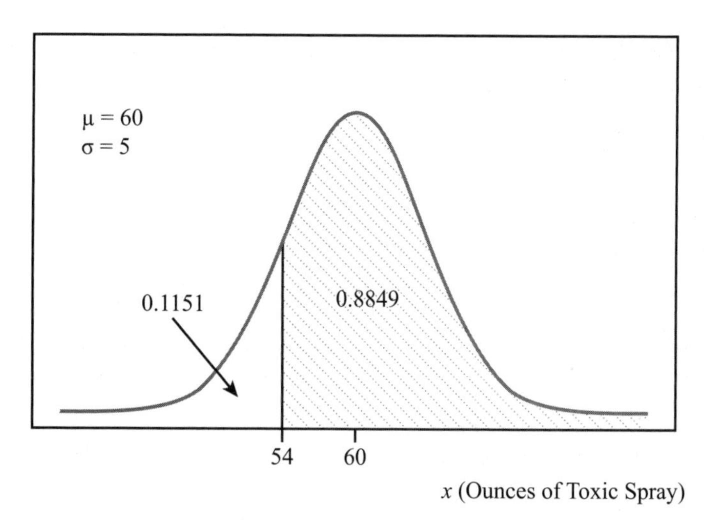

Figure 11.9
The shaded area is the probability that x will be more than 54 ounces.

Finally, let's look at the probability that Bob uses between 54 and 62.5 ounces of spray the next year. This probability is shown graphically in Figure 11.10.

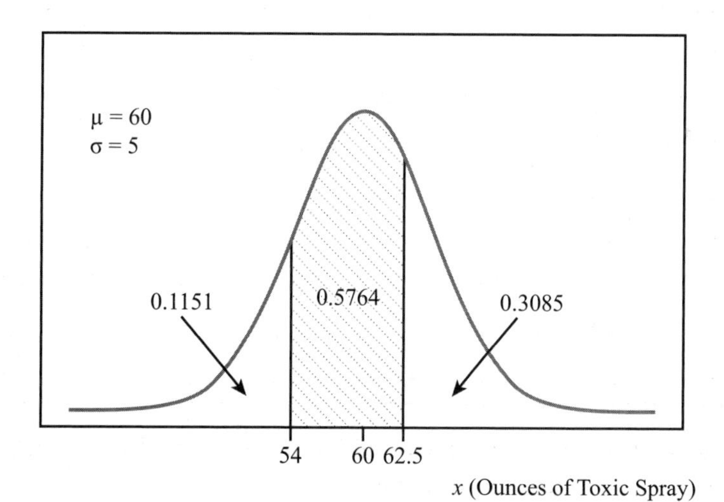

Figure 11.10
The shaded area is the probability that x will be between 54 and 62.5 ounces.

We know from previous examples that the area to the left of 54 ounces is 0.1151 and that the area to the right of 62.5 ounces is 0.3085. Because the total area under the curve is 1:

$$P(54 \leq x \leq 62.5) = 1 - 0.1151 - 0.3085 = 0.5764$$

There is a 57.64 percent chance that Bob uses between 54 and 62.5 ounces of spray the next year.

The Area Under the Normal Distribution and the Empirical Rule

The empirical rule (sounds like a decree from the emperor) tells us the area under the normal distribution. It states that if a distribution follows a bell-shaped, symmetrical curve centered around the mean, then we can expect approximately 68.3, 95.5, and 99.7 percent of the values to fall within 1.0, 2.0, and 3.0 standard deviations around the mean, respectively. I'm glad to inform you that we now have the ability to demonstrate these results.

The shaded area in Figure 11.11 shows the percentage of observations that we would expect to fall within 1.0 standard deviation of the mean.

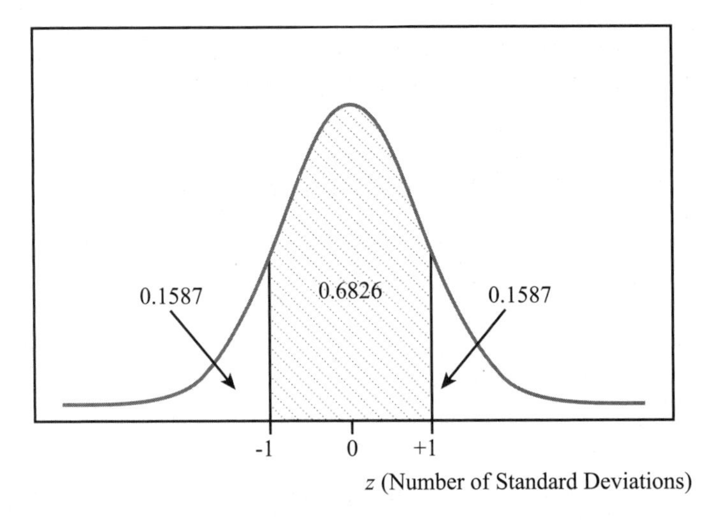

Figure 11.11
The shaded area is the probability that x will be between -1.0 and +1.0 standard deviation from the mean.

Where did 68.3 percent come from? We can look in the standard normal table to get the probability that an observation will be less than one standard deviation from the mean. The probability of the area on the right side ($z = +1$) is this:

$$P(z \leq +1.0) = 0.8413$$

And the probability of the area on the left side ($z = -1$) is this:

$$P(z < -1.0) = 0.1587$$

So the area between -1.0 and +1.0 is:

$$P(-1.0 \leq z \leq +1.0) = 0.8413 - 0.1587 = 0.6826$$

The same logic is used to demonstrate the probabilities of 2.0 and 3.0 standard deviations from the mean. I'll leave those for you to try.

The empirical rule is also known as the 68-95-99.7 percent rule. I'm sure you are not surprised where the name comes from!

Calculating Normal Probabilities Using Excel

Once again we can rely on Excel to do some of the grunt work for us. Excel has a built-in function, NORM.DIST, that can calculate the normal probability for us. It has the following characteristics:

NORM.DIST(*x*, mean, standard_dev, cumulative)

where:

cumulative = FALSE if you want the probability mass function (we don't)

cumulative = TRUE if you want the cumulative probability (we do)

For instance, Figure 11.12 shows the NORM.DIST function being used to calculate the probability that Bob uses less than 64.3 ounces of spray on those nasty beetles the next year.

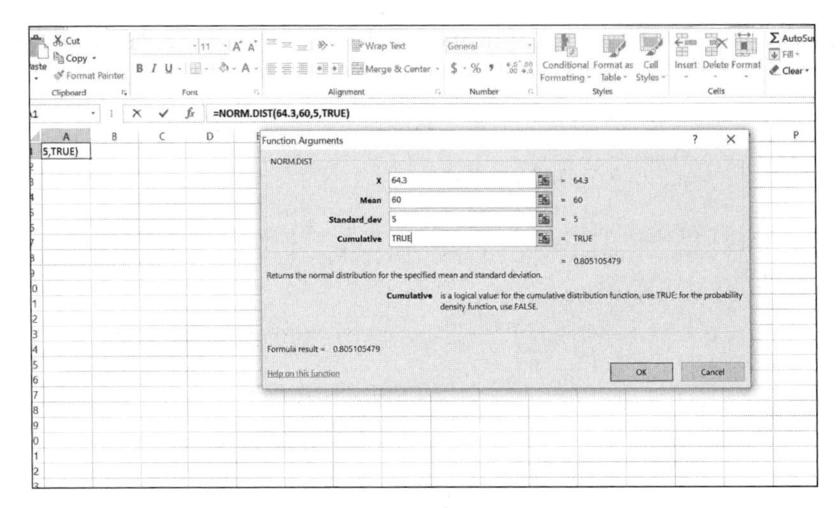

Figure 11.12

NORM.DIST function in Excel for less than 64.3 ounces.

Cell A1 contains the Excel formula =NORM.DIST (64.3,60,5,TRUE) with the result being 0.8051. This probability is underlined in the standard normal table earlier in the chapter.

BOB'S BASICS

Don't be alarmed if the values that are returned using the NORM.DIST function in Excel are slightly different than those found in Table 3 in Appendix B. This is due to rounding differences that are small enough to be ignored.

Using the Normal Distribution as an Approximation to the Binomial Distribution

Remember how nasty our friend the binomial distribution can get sometimes? Well, the normal distribution may be able help us out during these difficult times under the right conditions. Recall from Chapter 9 that the binomial equation will calculate the probability of x successes in n trials with p = the probability of a success for each trial and q = the probability of a failure. If $np \geq 5$ and $nq \geq 5$, then we can use the normal distribution to approximate the binomial.

As an example, suppose my statistics class is composed of 60 percent females. If I select 15 students at random, what is the probability that this group will include 8, 9, 10, or 11 female students? For this example, $n = 15$; $p = 0.6$; $q = 0.4$; and $x = 8, 9, 10$, and 11. We can use the normal approximation because $np = (15)(0.6) = 9$ and $nq = (15)(0.4) = 6$. (Sorry, guys. I didn't mean to infer picking you would be classified a failure!)

Also recall from Chapter 9 that the mean and standard deviation of this binomial distribution is this:

$$\mu = np = (15)(0.6) = 9$$

$$\sigma = \sqrt{npq} = \sqrt{(15)(0.6)(0.4)} = 1.897$$

The probability that the group of 15 students will include 8, 9, 10, or 11 female students can be calculated using the following equations:

$$P(x=8) = \frac{15!}{8!(15-8)!}(0.6)^8(0.4)^{15-8} = (6435)(0.0168)(0.0016) = 0.1730$$

$$P(x=9) = \frac{15!}{9!(15-9)!}(0.6)^9(0.4)^{15-9} = (5005)(0.0101)(0.0041) = 0.2073$$

$$P(x=10) = \frac{15!}{10!(15-10)!}(0.6)^{10}(0.4)^{15-10} = (3003)(0.0060)(0.0102) = 0.1838$$

$$P(x=11) = \frac{15!}{11!(15-11)!}(0.6)^{11}(0.4)^{15-11} = (1365)(0.0036)(0.0256) = 0.1258$$

$$P(x = 8, 9, 10, \text{ or } 11) = 0.1730 + 0.2073 + 0.1838 + 0.1258 = 0.6899$$

Now let's solve this problem using the normal distribution and compare the results. Figure 11.13 shows the normal distribution with $\mu = 9$ and $\sigma = 1.897$.

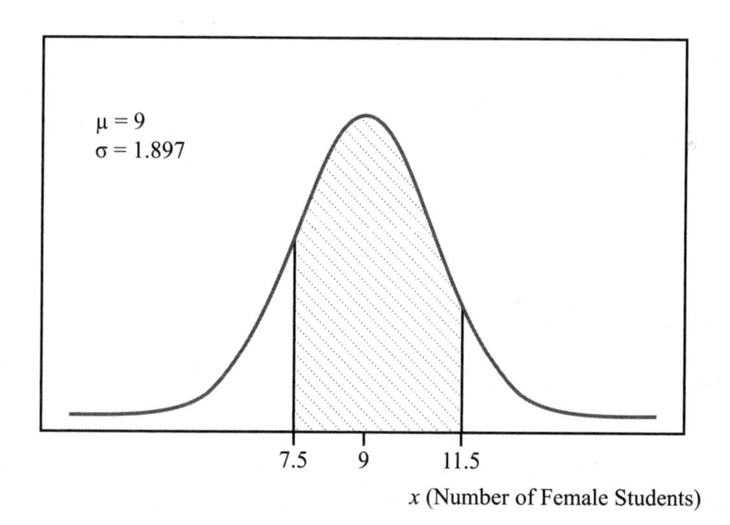

Figure 11.13
The normal approximation to the binomial distribution.

Notice that the shaded interval goes from 7.5 to 11.5 rather than 8 to 11. Don't worry; I didn't make a mistake. I subtracted 0.5 from 8 and added 0.5 to 11 to compensate for the fact that the normal distribution is continuous and the binomial is discrete. Adding and subtracting 0.5 is known as the continuity correction factor. For larger values of *n*, like 100 or more, you can ignore this correction factor.

Now we need to calculate the *z*-scores.

$$z = \frac{x-\mu}{\sigma} = \frac{11.5-9}{1.897} = +1.32$$

$$z = \frac{x-\mu}{\sigma} = \frac{7.5-9}{1.897} = -0.79$$

According to the standard normal table:

$P(z \leq +1.32) = .9066$

And $P(z \leq -0.79) = 0.2148$

The probability of interest for this example is the area between *z*-scores of -0.79 and +1.32. We can use the following calculations to find this area:

$P(-.079 \leq z \leq +1.32) = 0.9066 - 0.2148 = 0.6918$

This probability is shown in the shaded area in Figure 11.14.

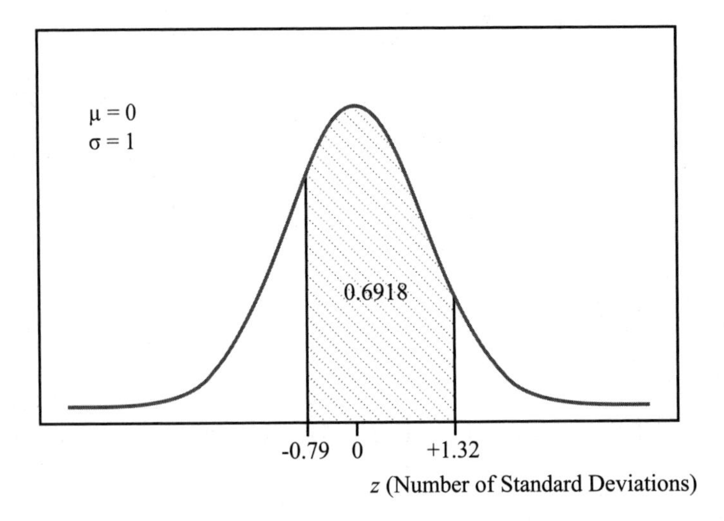

Figure 11.14
The probability that $-0.79 \leq z \leq +1.32$ *standard deviations from the mean.*

Using the normal distribution, we have determined the probability that my group of 15 students will contain 8, 9, 10, or 11 females is 0.6918. As you can see, this probability is very close to the result we obtained using the binomial equations, which was 0.6899.

Well, this ends our chapter on the normal probability distribution. Now we are prepared to dig deeper into inferential statistics!

Practice Problems

1. The speed of cars passing through a checkpoint follows a normal distribution with $\mu = 62.6$ miles per hour and $\sigma = 3.7$ miles per hour. What is the probability that the next car passing will …

   a. Be exceeding 65.5 miles per hour?

   b. Be exceeding 58.1 miles per hour?

   c. Be between 61 and 70 miles per hour?

2. The selling price of various homes in a community follows the normal distribution with $\mu = \$176,000$ and $\sigma = \$22,300$. What is the probability that the next house will sell for …

   a. Less than $190,000?

   b. Less than $158,000?

   c. Between $150,000 and $168,000?

3. The age of customers for a particular retail store follows a normal distribution with $\mu = 37.5$ years and $\sigma = 7.6$ years. What is the probability that the next customer who enters the store will be …

   a. More than 31 years old?

   b. Less than 42 years old?

   c. Between 40 and 45 years old?

4. A coin is flipped 14 times. Use the normal approximation to the binomial distribution to calculate the probability of a total of 4, 5, or 6 heads. Compare this to the binomial probability.

5. A certain statistics author's golf scores follow the normal distribution with a mean of 92 and a standard deviation of 4. What is the probability that, during his next round of golf, his score will be …

   a. More than 97?

   b. More than 90?

6. The number of text messages that Debbie's son Jeff sends and receives a month follows the normal distribution with a mean of 4,580 (I am not making this up!) and a standard deviation of 550. What is the probability that next month he will send and receive …

   a. Between 4,000 and 5,000 text messages?

   b. Less than 4,200 text messages?

7. A data set that follows a bell-shape and symmetrical distribution has a mean equal to 75 and a standard deviation equal to 10. What range of values centered around the mean would represent 95.5 percent of the data points?

The Least You Need to Know

- The normal distribution is bell-shaped and symmetrical around the mean.
- The total area under the normal distribution curve is equal to 1.0.
- The normal distribution tables are based on the standard normal distribution where $\mu = 0$ and $\sigma = 1$.
- The number of standard deviations between a normally distributed random variable, x, and μ is known as the standard z-score and can be found with $z = \frac{x-\mu}{\sigma}$.
- The empirical rule states that if a distribution follows a bell-shape, a symmetrical curve centered around the mean, then we can expect approximately 68.3, 95.5, and 99.7 percent of the values to fall within one, two, and three standard deviations around the mean, respectively.
- Excel has a built-in function, NORM.DIST, that you can use to perform normal distribution calculations.
- You can use the normal distribution to approximate the binomial distribution when $np \geq 5$ and $nq \geq 5$.

PART

3

Inferential Statistics

Now we can take all those wonderful concepts that we have stuffed into our overloaded brains from Parts 1 and 2 and put them to work using statistically sounding words, such as *confidence interval* and *hypothesis testing*. Inferential statistics enables us to make statements about a general population using the results of a random sample from that population. For instance, using inferential statistics, the winner of a political election can be accurately predicted very early in the polling process based on the results of a relatively small random sample that is properly chosen. Pretty cool stuff!

CHAPTER

12

Sampling

This first chapter dealing with the long-awaited topic of inferential statistics focuses on the subject of sampling. Way back in Chapter 1, we defined a population as representing all possible outcomes or measurements of interest, and a sample as a subset of a population. Here we'll talk about why we use samples in statistics and what can go wrong if they are not used properly.

Virtually all statistical results are based on the measurements of a sample drawn from a population. Major decisions are often made based on information from samples. For instance, the Nielson ratings gather information from a small sample of homes and are used to infer the television-viewing patterns of the entire country. The future of your favorite TV show rests in the hands of these select few! So choosing the proper sample is a critical step to ensure accurate statistical conclusions.

In This Chapter

- The reason for measuring a sample rather than the population
- The various methods for collecting a random sample
- Defining sampling errors and sampling bias
- Consequences for poor sampling techniques

Why Sample?

Most statistical studies are based on a sample of the population at large. The relationship between a population and sample is shown in Figure 12.1 (and also described in Chapter 1).

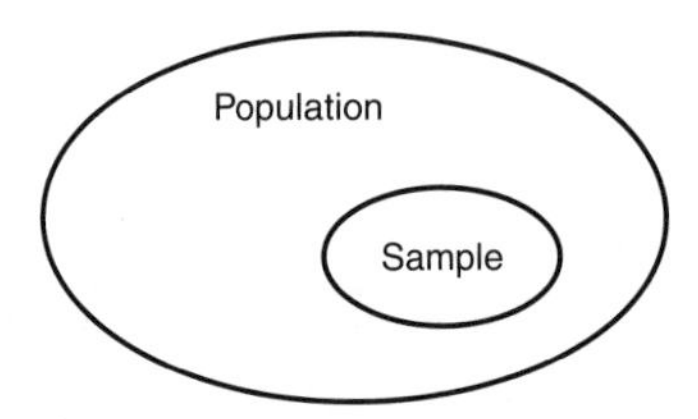

Figure 12.1
The relationship between a population and sample.

Why not just measure the entire population rather than rely on only a sample? That's a very good question! Depending on the study, measuring an entire population could be very expensive or just plain impossible. If I want to measure the life span of a certain breed of pesky mosquitoes (extremely short if I have any say in the matter), I could not possibly observe every single mosquito in the population. Rather, I would need to rely on a sample of the mosquito population, measure their life spans, and make a statement about the life span of the entire population. That's the whole concept of inferential statistics in one paragraph! Unfortunately, doing what we just wrote is a whole lot harder than just writing it. Doing it is what the rest of this book is all about!

Even if we could feasibly measure the entire population, to do so would often be a wasteful decision. If a sample is collected properly and the analysis performed correctly, we can make a very accurate assessment of the entire population. There is very little added benefit to continue beyond the sample and measure everything in sight. Measuring the population often is a waste of both time and money, resources that seem to be very scarce these days.

One example where such a decision was recently made occurred at Goldey-Beacom College where we teach. Bob was the Chair of the Academic Honor Code Committee and was involved in a project whose goal was to gather information regarding the attitude of our student body on the topic of academic integrity. It would have been possible to ask every student at our college to respond to the survey, but it was really unnecessary with the availability of inferential statistics. We eventually made the intelligent decision and sampled only a portion of the students to infer the attitudes of the population.

BOB'S BASICS

Often it is just not feasible to measure an entire population. Even when it is feasible, measuring an entire population can be a waste of time and money and provides little added benefit beyond measuring a sample.

Random Sampling

The term *random sampling* refers to a sampling procedure where every member in the population has a chance of being selected. The objective of the sampling procedure is to ensure that the final sample to be measured is representative of the population from which it was taken. If this is not the case, then we have a *biased sample*, which can lead to misleading results. If you recall, we discussed an example of a biased sample back in Chapter 1 with the golf course survey. The selection of a proper sample is critical to the accuracy of the statistical analysis.

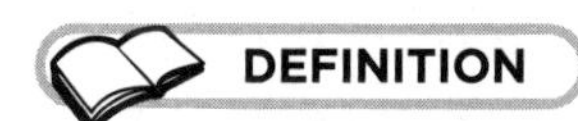

DEFINITION

Random sampling refers to a sampling procedure where every member in the population has a chance of being selected. A **biased sample** is a sample that does not represent the intended population and can lead to distorted findings.

As we will see in the following sections, there are four different ways to gather a random sample: simple random, systematic, cluster, and stratified.

Simple Random Sampling

A *simple random sample* is a sample in which every member of the population has an equal chance of being chosen. Unfortunately, this is easier said than done. To illustrate, let's use the academic integrity example mentioned previously.

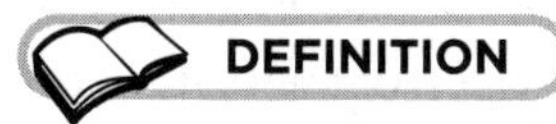

DEFINITION

A **simple random sample** is a sample in which every member of the population has an equal chance of being chosen.

There are a few options for choosing a simple random sample for the academic integrity survey example. I could randomly choose students using a *random number table*, which is aptly named. (After all, it is simply a table of numbers that are completely random.) An excerpt of such a table is shown here:

57245	39666	18545	50534	57654	25519	35477	71309	12212	98911
42726	58321	59267	72742	53968	63679	54095	56563	09820	86291
82768	32694	62828	19097	09877	32093	23518	08654	64815	19894
97742	58918	33317	34192	06286	39824	74264	01941	95810	26247
48332	38634	20510	09198	56256	04431	22753	20944	95319	29515
26700	40484	28341	25428	08806	98858	04816	16317	94928	05512
66156	16407	57395	86230	47495	13908	97015	58225	82255	01956
64062	10061	01923	29260	32771	71002	58132	58646	69089	63694
24713	95591	26970	37647	26282	89759	69034	55281	64853	50837
90417	18344	22436	77006	87841	94322	45526	38145	86554	42733

Suppose we had 1,000 students in the population from which we were drawing a sample size of 100. (We'll discuss sample size in Chapter 14.) I would list these students with assigned numbers from 0 to 999. Looking at the first three digits in each cell, the random number table would tell me to select student 572, followed by student 427, and so forth until I had selected 100 students. Using this technique, my sample of 100 students would be chosen with complete randomness.

Another method to choose a simple random sample is to use the Sampling function in Excel, part of the Analysis ToolPak. Using the academic integrity survey example, let's say I want to choose 20 students at random for the sample using the ID numbers out of 200 students. I put the GBC student ID numbers in Column A. I'll follow these steps:

1. Go to the Data tab, then click on Data Analysis.
2. From the Data Analysis box, select Sampling, and click OK.
3. In the Input Range box, select the data in column A (cells A1 to A201 in this example). Check the Labels box because we have a label in cell A1.
4. Under Sampling Method, choose Random and type "20" in the Number of Samples box.
5. In the Output Range box, choose the cell where you want Excel to put your randomly chosen sample (cell D2 in this example), and click OK.

Figure 12.2 shows the Excel Sampling function described above.

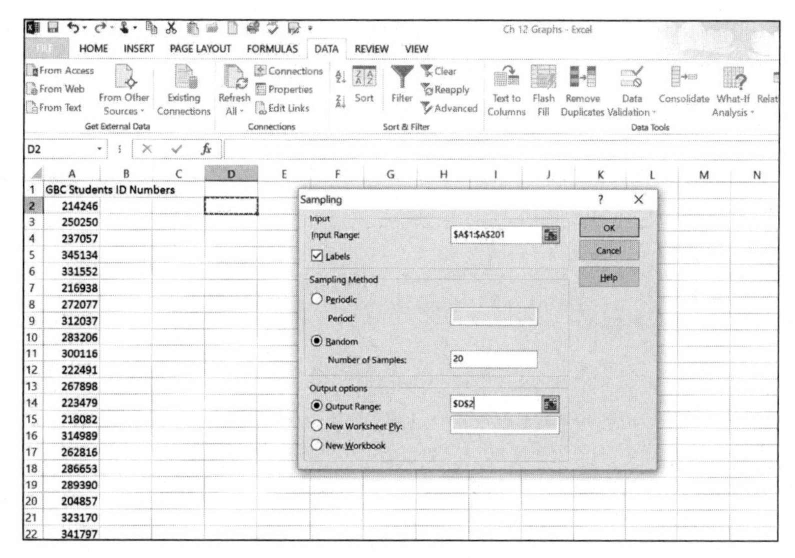

Figure 12.2
Excel's Sampling function.

The result of the 20 random sample students is displayed in cells D2 to D21, as shown in Figure 12.3.

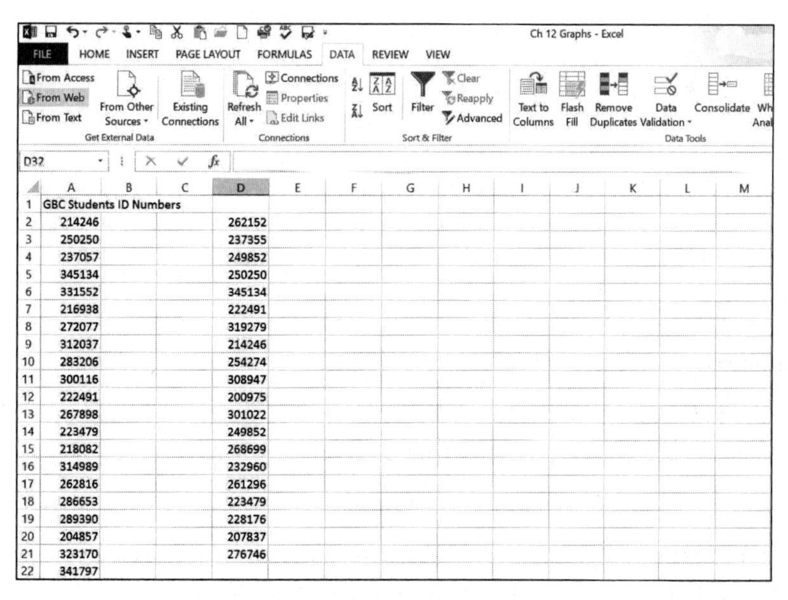

Figure 12.3
A random sample of 20 student ID Numbers.

So the first chosen student in the sample is the one with ID number 262152, followed by the student with ID number 237355, and so forth. All students had an equal chance of being chosen. Note that Excel uses sampling *with replacement*, meaning that after the first student ID number is chosen, it's put back into the population before the second student ID number is chosen. Therefore, Excel can choose the same ID number more than once. If you don't want the same student to be chosen more than once, then replace the duplicated student in your sample with another one at random.

Systematic Sampling

One way to avoid a personal bias when selecting people at random is to use *systematic sampling.* This technique results in selecting every k^{th} member of the population to be in your sample. The value of k will depend on the size of the sample and the size of the population. Using our academic integrity survey, with a population of 1,000 students and a sample of 100, $k = 10$. From a listing of the entire population, I would choose every tenth student to be included in the sample. In general, if N = the size of the population and n = the size of the sample, the value of k would be approximately $\frac{N}{n}$.

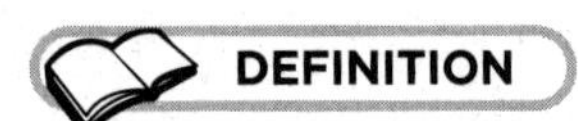

In **systematic sampling**, every k^{th} member of the population is chosen for the sample, with the value of k being approximately $\frac{N}{n}$.

You can also use Excel for systematic sampling. Just follow the same steps as with the simple random sample before, except you want Periodic this time when you choose the sampling method. In the Period box, type your k value, which is 10 in this example (shown in Figure 12.4). To find k, we used the equation $\frac{N}{n}$, where $N = 200$ and $n = 20$.

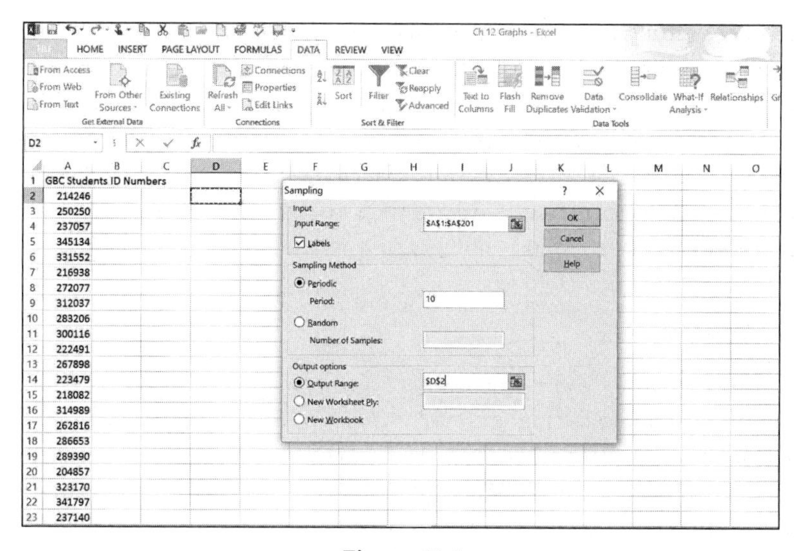

Figure 12.4
Systematic sampling in Excel.

The chosen systematic sample is placed in cells D2 to D21, as shown in Figure 12.5. This sample includes the 10th student on the list, followed by the 20th student on the list, and so forth.

	A	B	C	D
1	GBC Students ID Numbers			
2	214246			300116
3	250250			323170
4	237057			331909
5	345134			293776
6	331552			238220
7	216938			237730
8	272077			240774
9	312037			281297
10	283206			227824
11	300116			204523
12	222491			334582
13	267898			242683
14	223479			326228
15	218082			233308
16	314989			338267
17	262816			301022
18	286653			203891
19	289390			347345
20	204857			223035
21	323170			337695
22	341797			

Figure 12.5
A systematic sample of 20 Students.

The benefit of systematic sampling is that it's easier to conduct than a simple random sample, often resulting in less time and money. The downside is the danger of selecting a biased sample if there is a pattern in the population that is consistent with the value of k. For instance, let's say I'm conducting a survey on campus asking students how many hours they are studying during the week, and I select every fourth week to collect my data. Because we are on an 8-week semester schedule at Goldey-Beacom College, every fourth week could end up being mid-terms and finals week, which would result in a higher number of study hours than normal (or at least I would hope so!).

Cluster Sampling

If we can divide the population into groups, or *clusters*, then we can select a simple random sample from these clusters to form the final sample. Using the academic integrity survey example, the clusters could be defined as classes. We would randomly choose different classes to participate in the survey. In each class chosen, every student would be selected to be part of the sample.

A **cluster** sample is a simple random sample of groups, or clusters, of the population. Each member of the chosen clusters would be part of the final sample.

For cluster sampling to be effective, it is assumed that each cluster selected for the sample is representative of the population at large. In effect, each cluster is a miniaturized version of the overall population. If used properly, cluster sampling can be a very cost-effective way of collecting a random sample from the population.

Stratified Sampling

In *stratified sampling*, we divide the population into mutually exclusive groups, or strata, that have something in common, and we randomly sample from each of these groups. There are many different ways to establish strata from the population. Using the academic integrity survey, we could define our strata as undergraduate and graduate students. If 20 percent of our college population is graduate students, then I could use stratified sampling to ensure that 20 percent of my final sample is also composed of graduate students. Other examples of criteria that we can use to divide the population into strata are age, income, or occupation. Stratified sampling is helpful when it is important that the final sample has certain characteristics of the overall population.

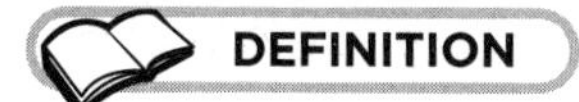

A **stratified sample** is obtained by dividing the population into mutually exclusive groups, or strata, with a common characteristic and then randomly sampling from each of these groups.

One difference between stratified and cluster sampling is in the choice of the groups. With stratified sampling, the strata have something in common, such as having graduate or undergraduate status in our example. However, in cluster sampling, the strata are sub-sets of the population and don't necessarily have to have something in common.

Sampling Error and Sampling Bias

Up to this point, we have stressed the benefits of drawing a sample from a population rather than measuring every member of the population. However, in statistics, as in life, there's no such thing as a free lunch. By relying on a sample, we expose ourselves to errors that can lead to inaccurate conclusions about the population.

Sampling Error

The type of error that a statistician is most concerned about is called *sampling error*, which occurs when the sample measurement is different from the population measurement. Because the population is rarely measured in its entirety, the sampling error cannot be directly calculated. However, with inferential statistics, we'll be able to assign probabilities to certain amounts of sampling error later in Chapter 15.

Sampling error results from the difference between the population parameter (data which describes something about the population) and the sample statistic (data which describes the sample). Because samples don't perfectly represent the population, we get sampling errors.

Sampling errors occur because we might have the unfortunate luck of selecting a sample that is not a perfect match to the entire population. Sampling errors are expected and usually are a small price to pay to avoid measuring an entire population. One way to reduce the sampling error of a statistical study is to increase the size of the sample. In general, the larger the sample size, the smaller the sampling error. If you increase the sample size until it reaches the size of the population, then the sampling error will be reduced to zero. But in doing so, you forfeit the benefits of sampling.

Sampling Bias

Sampling bias occurs when samples are designed to choose some of the population only with certain characteristics instead of parts of the population without those characteristics. For example, let's say I want to show that the current administration is doing a good job handling the economy by showing that the average income in the United States is high. In creating my sample, I choose more people from the states with high income and less people from the states with low income. This way, the average for my sample will show a high average income in the United States. In this case, my sample is biased. It shows incorrectly that average income in the United States is higher than it actually is.

To demonstrate sampling bias in another instance, I might want to show that the unemployment rate in the United States is low. In designing my sample, I choose more people from the states with low unemployment rates and less people from the states with high unemployment rates. This way my sample will show a low unemployment rate in the United States. Again, this sample is biased and incorrectly represents the population.

To avoid sampling bias, samples should be chosen to represent the entire population from which they are drawn and every member in the population should have the same chance of being selected. In other words, be careful that your sample does not just include parts of the population with certain characteristics over parts without those characteristics.

Examples of Poor Sampling Techniques

The technique of sampling has been widely used, both properly and improperly, in the area of politics. One of the most famous mishaps with sampling occurred during the 1936 presidential race when the *Literary Digest* predicted Alf Landon to win the election over Franklin D. Roosevelt. Even if history is not your best subject, you can realize somebody had egg on his face after this election day. *Literary Digest* drew their sample from phonebooks and automobile registrations. The problem was that people with phones and cars in 1936 tended to be wealthier Republicans and were not representative of the entire voting population.

Another sampling blunder occurred in the 1948 presidential race when the Gallup poll predicted Thomas Dewey to be the winner over Harry Truman. The failure of the Gallup poll stemmed from the fact that there were a large number of undecided voters in the sample. It was wrongly assumed that these voters were representative of the decided voters who happened to favor Dewey. Truman easily won the election with 303 electoral votes compared to Dewey's 189.

As you can see, choosing the proper sample is a critical step when using inferential statistics. Even a large sample size cannot hide the errors of choosing a sample that is not representative of the population at large. History has shown that large sample sizes are not needed to ensure accuracy. For example, the Gallup poll predicted that Richard Nixon would receive 43 percent of the votes for the 1968 presidential election and in fact he won 42.9 percent. This Gallup poll was based on a sample size of only 2,000; whereas the disastrous 1936 *Literary Digest* poll sampled 2,000,000 people (Source: personal.psu.edu/faculty/g/e/gec7/Sampling.html).

WRONG NUMBER

Have you ever participated in an online survey on a sports or news website that allowed you to view the results? These surveys can be fun and interesting, but you need to take the results with a grain of salt. That's because the respondents are self-selected, which means the sample is *not* randomly chosen. The results of these surveys are most likely biased because the respondents would not be representative of the population at large. For example, people without internet access would not be part of the sample and might respond differently than people with access to the internet.

Practice Problems

1. You are to gather a systematic sample from a local phone book with 75,000 names. If every k^{th} name in the phone book is to be selected, what value of k would you choose to gather a sample size of 500?
2. Consider a population that is defined as every employee in a particular company. How could you use cluster sampling to gather a sample to participate in a survey involving employee satisfaction?
3. Consider a population that is defined as every employee in a particular company. How could you use stratified sampling to gather a sample to participate in a survey involving employee satisfaction?

The Least You Need to Know

- A simple random sample is a sample in which every member of the population has an equal chance of being chosen.
- In systematic sampling, every kth member of the population is chosen for the sample, with the value of k being approximately $\frac{N}{n}$.
- A cluster sample is a simple random sample of groups, or clusters, of the population. Each member of the chosen clusters would be part of the final sample.

- Obtain a stratified sample by dividing the population into mutually exclusive groups, or strata, and randomly sampling from each of these groups.
- Sampling error occurs when the sample measurement is different from the population measurement. It is the result of selecting a sample that is not a perfect match of the entire population.

CHAPTER

13

Sampling Distributions

In Chapter 12, we praised the wonders of using samples in our statistical analysis because it was more efficient than measuring an entire population. In this chapter, we'll discover another benefit of using samples—sampling distributions.

Sampling distributions describe how sample averages behave. You may be surprised to hear they behave very well—even better than the populations from which they are drawn. Good behavior means we can do a pretty good job at predicting future values of sample means with a little bit of information. This might sound a little puzzling now, but by the end of this chapter you'll be shaking your head in utter amazement.

In This Chapter

- Using sampling distributions of the mean and proportion
- Working with the central limit theorem
- Using the standard error of the sampling mean and proportion

What Is a Sampling Distribution?

The sampling distribution is a table with two columns: each of the sample statistics (such as the sample means) in one column and the corresponding probabilities in the other column. It is very similar to the probability distribution we talked about in Chapter 8, with the difference being the variable. In this case, we record the sample statistic instead of the values of the random variable, like before.

For example, let's say I want to perform a study to determine the number of miles the average person drives a car in one day. Because it's not possible to measure the driving patterns of every person in the population, I randomly choose a sample size of 10 ($n = 10$) qualified individuals and record how many miles they drove yesterday. I then choose another 10 drivers and record the same information. I do this three more times, with the results in the following table.

Sample Number	Average Number of Miles (Sample Mean)
1	40.4
2	76.0
3	58.9
4	43.6
5	62.6

As you can see, each sample has its own mean value, and each value is different. We can continue this experiment by selecting many more samples and observing the pattern of sample means. This pattern of sample means represents the sampling distribution for the number of miles the average person drives in one day.

Sampling Distribution of the Sample Means

The distribution from the previous example represents the *sampling distribution of the sample means* because the mean of each sample was the measurement of interest. Why do we care about the sampling distribution? Good question—we care because it has interesting properties that help us in inferential statistics. If we sample with replacement from a population and we get all possible samples of a given size (Do you remember counting rules? This is the method we would use to get all the possible samples of certain size from a population.), then the sampling distribution will have these cool features:

- The mean of this sampling distribution will have the exact mean as the population. Amazing, isn't it? But wait to see the next one.

- The sampling distribution of the sample mean will have a smaller standard deviation than the population from which the samples were drawn. In other words, the sampling distribution is narrower than the distribution for the population. Moreover, the standard deviation of this sampling distribution equals the population standard deviation divided by the square root of the sample size, *n*.

Let's explore these features in detail and use an example to clarify. Just for simplicity and to make the calculations easier, let's say I've a small class of five students with the following grades on a statistics exam (so my entire population consists of these 5 grades):
90 88 86 84 82. Now I want to get all possible samples (of say 2 students) with replacement from this class. To do that, we use the fundamental counting rule from Chapter 7, and we get 25 samples. The following table lists all 25 sample combinations of 2 students and the corresponding sample mean of each one.

Sample		**Sample Mean ($\bar{x}$)**
90	90	90
90	88	89
90	86	88
90	84	87
90	82	86
88	90	89
88	88	88
88	86	87
88	84	86
88	82	85
86	90	88
86	88	87
86	86	86
86	84	85
86	82	84
84	90	87
84	88	86
84	86	85

continues

continued

Sample		Sample Mean ($\bar{x}$)
84	84	84
84	82	83
82	90	86
82	88	85
82	86	84
82	84	83
82	82	82

Let's check those two cool features:

First we'll calculate the means. The population mean = $\frac{90+88+86+84+82}{5}$ = 86. The mean of all the sample means in the last column of the table is $\frac{90+89+88+87+.....+84+83+82}{25} = \frac{2150}{25}$ = 86, also!

If you're not surprised yet, look at this second feature! Recall the equation for finding the population standard deviation: $\sigma = \sqrt{\frac{\sum_{i=1}^{N}(\bar{x}_i - \mu)^2}{N}}$. Using this, the standard deviation of the population = 2.83.

The standard deviation of all the sample means in the last column, using $\sigma = \sqrt{\frac{\sum_{i=1}^{N}(\bar{x}_i - \mu)^2}{N}}$, is 2.00.

Look at this: divide the standard deviation of the population by the $\sqrt{n}$ and you get $\frac{2.83}{\sqrt{2}}$ = 2.00, which we know is also the standard deviation of the sample means! Ta da! Please hold your applause until the end of the book!

Now you know why we statisticians are very fond of sampling distributions!

The **sampling distribution of the sample means** is a table with two columns: the sample means in one column and the probability of each sample mean in another column.

Mean of the Sampling Distribution of the Sample Means

Using our previous example, we can get the sampling distribution of the sample means by listing all the sample means in one column and the probability of each one in the second column, as follows:

Sampling Distribution of the Sample Mean ($n = 2$)

Sample Mean ($\bar{x}$)	Frequency	Probability (Relative Frequency)
82	1	0.04
83	2	0.08
84	3	0.12
85	4	0.16
86	5	0.20
87	4	0.16
88	3	0.12
89	2	0.08
90	1	0.04

We can also display this sampling distribution graphically, shown in Figure 13.1.

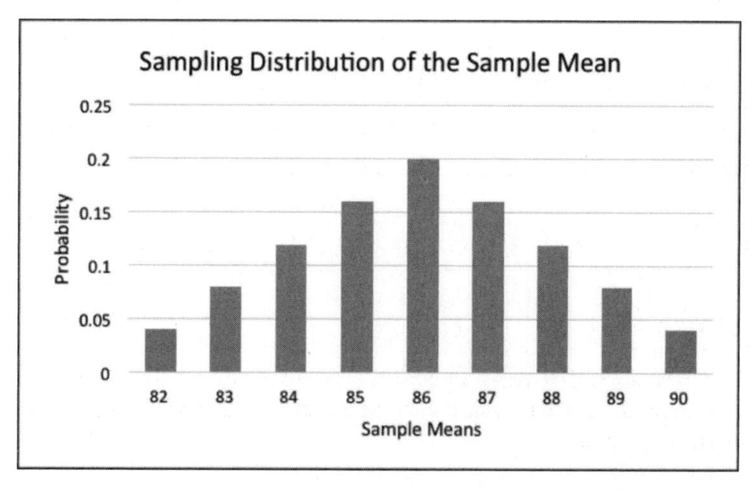

Figure 13.1
Sample distribution of the sample means for $n = 2$.

Students often confuse sample size, n, and number of samples. In the previous example, the sample size equals 2 ($n = 2$), and the number of samples equals 25. In other words, we have 25 samples, each whose sample size is 2.

In the previous example we proved that the mean of the sampling distribution is the same as the population mean. But if you still doubt it, let's prove it another way. Applying the same formula for the mean of the probability distribution (which we learned in Chapter 8) to this sampling distribution gives us the following:

$$\mu_{\overline{X}} = \sum_{i=1}^{N} \overline{x}_i P(\overline{x}_i)$$

where:

$\mu_{\overline{X}}$ = the mean of the sampling distribution

$\overline{x}_i$ = the sample mean

$P(\overline{x}_i)$ = the probability of the sample mean

N = number of samples

The table below shows the calculation for the mean using the previous formula:

Sample Mean ($\overline{x}_i$)	$P(\overline{x}_i)$	$\overline{x}_i P(\overline{x}_i)$
82	0.04	3.28
83	0.08	6.64
84	0.12	10.08
85	0.16	13.6
86	0.20	17.2
87	0.16	13.92
88	0.12	10.56
89	0.08	7.12
90	0.04	3.6

$$\mu_{\overline{X}} = \sum_{i=1}^{N} \overline{x}_i P(\overline{x}_i) = 3.28 + 6.64 + 10.08 + 13.6 + 17.2 + 13.92 + 10.56 + 7.12 + 3.6 = 86$$

As you can see your answer is again 86. Now you'll always believe me! The mean of the sampling distribution of the sample means is the same as the population mean.

Standard Error of the Sampling Distribution of the Sample Means

In the previous example, we saw that the standard deviation of the sampling distribution equals $\frac{\sigma}{\sqrt{n}}$. Now to prove it to you in a different way, let's apply to this example the variance and standard deviation formulas for the probability distribution that we learned in Chapter 8:

$$\sigma_{\bar{X}}^2 = \sum_{i=1}^{N} (\bar{x}_i - \mu)^2 P(\bar{x}_i) \text{ and } \sigma_{\bar{X}} = \sqrt{\sigma_{\bar{X}}^2}$$

where:

$\sigma_{\bar{X}}$ = the standard deviation of the sampling distribution of the sample means

The following table shows the calculation for the standard deviation using the previous formula:

Sample Mean ($\bar{x}_i$)	$P(\bar{x}_i)$	$(\bar{x}_i - \mu)^2 P(\bar{x}_i)$
82	0.04	0.64
83	0.08	0.72
84	0.12	0.48
85	0.16	0.16
86	0.20	0
87	0.16	0.16
88	0.12	0.48
89	0.08	0.72
90	0.04	0.64

$\sigma_{\bar{X}}^2 = 0.64 + 0.72 + 0.48 + 0.16 + 0 + 0.16 + 0.48 + 0.72 + 0.64 = 4.00$

$\sigma_{\bar{X}} = \sqrt{4} = 2$ (which is equal to $\frac{\sigma}{\sqrt{n}} = \frac{2.83}{\sqrt{2}}$).

Just like the mean, the answer is the same as the one we obtained in the previous section. Therefore, we can calculate the standard deviation of the sampling distribution of the sample means as follows:

$$\sigma_{\bar{X}} = \frac{\sigma}{\sqrt{n}}$$

where:

$\sigma_{\bar{X}}$ = the standard deviation of the sampling distribution of the sample means

σ = the standard deviation of the population

n = sample size

Just to throw one more term at you, the standard deviation of the sampling distribution (what we just calculated) is also known as the *standard error of the mean.*

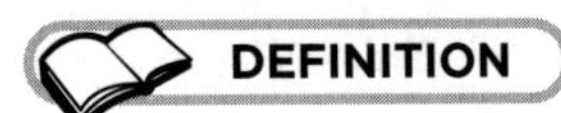

The **standard error of the mean** is the standard deviation of the sampling distribution of the sample means and can be determined by $\sigma_{\bar{X}} = \frac{\sigma}{\sqrt{n}}$.

BOB'S BASICS

Students often confuse σ and $\sigma_{\bar{X}}$. The symbol σ, the standard deviation of the population, measures the variation within the population and was discussed in Chapter 5. The symbol $\sigma_{\bar{X}}$, the standard error, measures the variation of the sample means and will decrease as the sample size increases.

I'm sure by now your highly inquisitive mind is screaming, "What happens to the sampling distribution if we increase the sample size?" That's an excellent question, which we will address in the next section.

The Central Limit Theorem

As we mentioned earlier, sample means behave in a very special way. According to the *central limit theorem*, as the sample size, *n*, gets larger, the sample means tend to follow a normal probability distribution. This holds true regardless of the distribution of the population from which the sample was drawn. Amazing, you say.

According to the **central limit theorem**, as the sample size, n, gets larger, the sampling distribution of the sample means tends to follow a normal probability distribution with a mean equal to the true population mean, μ, and standard error $\sigma_{\bar{X}} = \frac{\sigma}{\sqrt{n}}$. This holds true regardless of the distribution of the population from which the sample was drawn.

The central limit theorem is very important in statistics, and as Bob called it, it's "the mother of all theorems." The central limit theorem assures us that if we sample large enough ($n \geq 30$), then the sampling distribution will be normally distributed regardless of the distribution of the population itself. If the population from which the samples were drawn is not normal or if we simply don't know whether the population is normal or not, then the central limit theorem will hold as long as we have a large sample of 30 or more. Note, however, that if the population from which the samples were drawn is normally distributed, then the sampling distribution taken from it is also normally distributed without the need for the central limit theorem (in other words, it would work for any sample size).

Remember how we said the sampling distribution behaves very well? To show you an example of its good behavior, compare the probability distribution of the population to the probability distribution of the sampling distribution of the sample means. Since $\sigma_{\bar{X}} = \frac{\sigma}{\sqrt{n}}$ and $\sqrt{n} > 0$, then $\sigma_{\bar{X}} < \sigma$. You say, "So what?" Well, it means that the sampling distribution is less dispersed around the mean (in other words, closer to the mean) than the distribution for the population from which the samples were drawn. That's really good behavior, isn't it? This is clear in Figures 13.2 and 13.3. Figure 13.2 shows the normal distribution for the population in our example, whereas Figure 13.3 shows the normal distribution for the sampling distribution for our example. You can see the sampling distribution is skinnier than the population distribution! Moreover, as the sample size (n) increases, $\sigma_{\bar{X}}$ gets smaller so the sampling distribution gets even closer to the mean—it's getting skinnier!

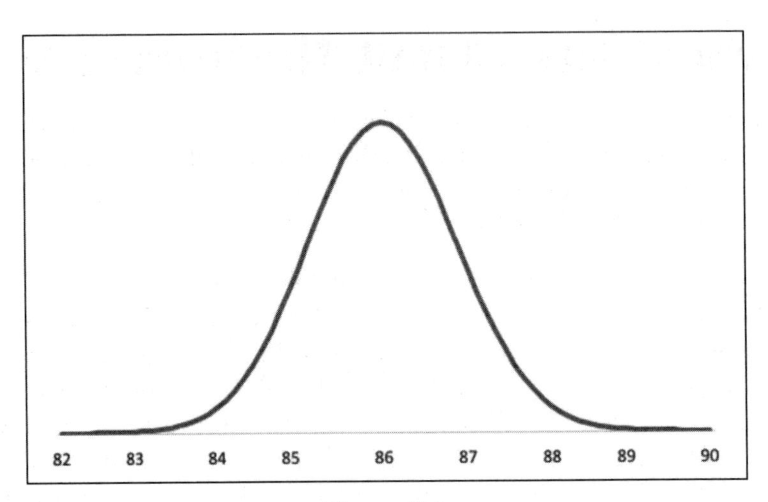

Figure 13.2
The normal distribution for the population in the grade example.

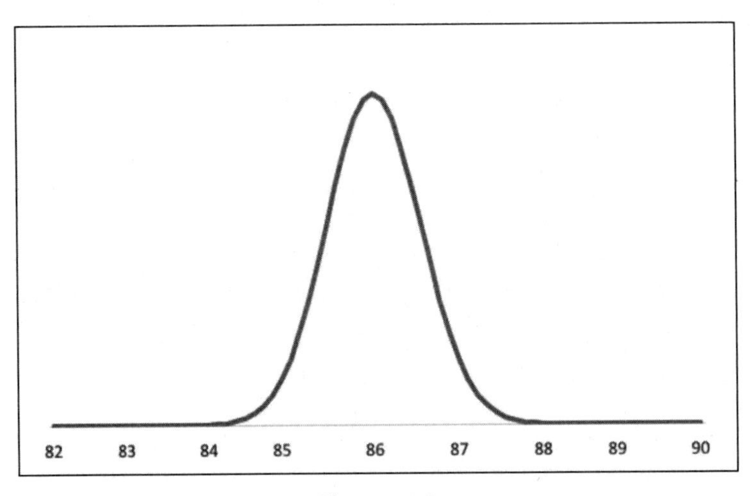

Figure 13.3
The normal distribution for the sampling distribution in the grade example.

Putting the Central Limit Theorem to Work

I can just sense your need right now to do something really neat with this wonderful new tool. Look no further. If we know the sampling distribution of the sample means follows the normal probability distribution and we also know the mean and standard deviation of that distribution, we can predict the likelihood that the sample means will be greater or less than certain values.

To clarify, let's look at an example. According to MyFICO.com, the average FICO score in the United States in October 2012 was 689. Because I don't know if FICO scores are normally distributed or not, I can't calculate the probability that the FICO score for any single individual is less than any specific value, say 695. Using the central limit theorem, however, I can calculate the probability that a randomly selected sample of say 30 individuals will have an average FICO score of less than 695. How can I do that? With my large sample, the central limit theorem assures me that the sampling distribution of the sample mean is normally distributed with:

$$\mu = 689$$

Now, we need to calculate $\sigma_{\bar{X}}$. Let's assume that the standard deviation for the FICO score is 15, then:

$$\sigma_{\bar{X}} = \frac{\sigma}{\sqrt{n}} = \frac{15}{\sqrt{30}} = 2.74$$

Knowing that the sampling distribution of the sample means follows a normal distribution, we can use the standard normal distribution table to calculate the probability. As we did in Chapter 11, we need to calculate the z-score. The equation looks slightly different because we are working with sample means, but in reality, it is identical to what we saw in Chapter 11. We just replace x with $\bar{x}$ and use its mean and standard deviation instead, as follows:

$$z = \frac{\bar{x} - \mu}{\sigma_{\bar{x}}}$$

$$z = \frac{695 - 689}{2.74} = 2.19$$

Using the standard normal z-table in Appendix B:

$$P(\bar{x} < 695) = P(z < 2.19) = 0.9857$$

This probability is shown in Figure 13.4.

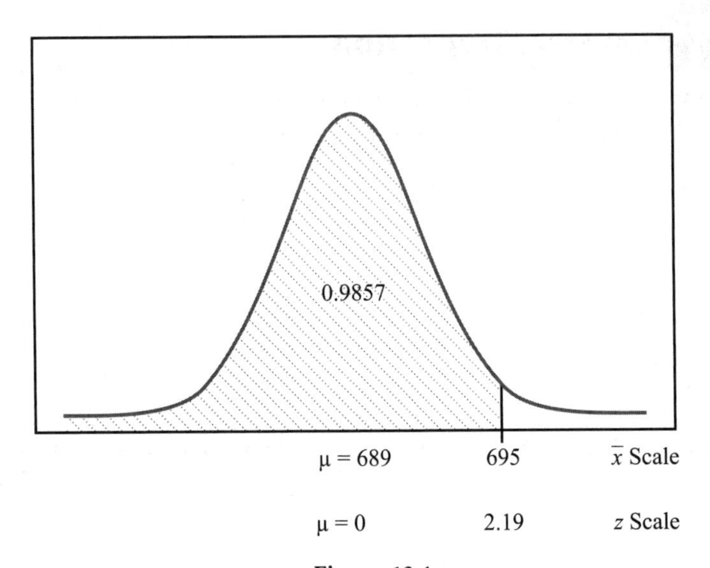

Figure 13.4

Probability that the sample mean for FICO score is less than 695.

According to the shaded area, the probability that the sample mean for the FICO score is less than 695 is 98.6 percent.

As you can see, the power of the central limit theorem lies in the fact that you need little information about the distribution of the population to apply it. The sample means will behave very nicely as long as the sample size is large enough. It's a very versatile theorem that has countless applications in the real world. I knew you'd be impressed!

BOB'S BASICS

The central limit theorem is one of the most powerful concepts for inferential statistics. It forms the foundation for many statistical models that are used today, so it's a good idea to cozy up to this theorem.

Sampling Distribution of the Proportion

The sample mean is not the only statistical measurement that is performed. What if I want to measure the sample proportion instead? For example, I want to know the percentage of working people who are satisfied with their job. So I collect a random sample of 300 workers and ask them whether they are satisfied with their job or not. Because each respondent has only two choices (satisfied or unsatisfied), this experiment follows the binomial probability distribution, which we discussed in Chapter 9. We also saw that we can use the normal distribution to approximate the binomial distribution and that's what we are going to do here. So let's look at it step by step.

Calculating the Sample Proportion

My measurement of interest is the proportion of workers in my sample of size n, who are satisfied with their job. The sample proportion, $\overline{p}$, is calculated by:

$$\overline{p} = \frac{x}{n}$$

where:

x: number of successes

n: sample size

Note that the sample proportion is denoted by $\overline{p}$, while the population proportion is denoted by p. According to the Conference Board, in 2013 47.7 percent of workers were satisfied with their job. Now, $p = 0.477$.

As you recall from Chapter 11, we can use the normal probability distribution to approximate the binomial distribution if $np \geq 5$ and $nq \geq 5$ ($q = 1 - p$, the probability of a failure). I can check both conditions as follows:

$$np = (300)(0.477) = 143.10$$

$$nq = (300)(1 - 0.477) = 156.90$$

WRONG NUMBER

It's important to remember that a proportion, either p or $\overline{p}$, cannot be less than 0 or greater than 1. A common mistake that students make is when told that the proportion equals 10 percent, they set $p = 10$ rather than $p = 0.10$.

Just like the sample mean, I can get the sampling distribution of the sample proportions by collecting different samples of the same size and getting the proportion of each sample. This sampling distribution of the proportion, just like the sampling distribution of the sample means, will follow the normal distribution and the average for these sample proportions, yes you guessed it, will be the same as the population proportion, p.

Calculating the Standard Error of the Proportion

I now need to calculate the standard deviation of this sampling distribution, which is known as the *standard error of the proportion*, or σ_p, with the following equation:

$$\sigma_p = \sqrt{\frac{p(1-P)}{n}}$$

where

p: the population proportion

n: sample size

In our example, $\sigma_p = \sqrt{\frac{0.477(1-0.477)}{300}} = 0.0288.$

The **standard error of the proportion** is the standard deviation of the sample proportions and can be calculated by $\sigma_p = \sqrt{\frac{p(1-P)}{n}}$.

We're now ready to use this useful information to answer questions like, "What is the probability that 130 workers or more in my sample are satisfied with their job?"

To calculate the probability, we need to calculate the z-value for the proportion using the following equation:

$$z = \frac{\bar{p}-p}{\sigma_p}$$

TEST YOUR KNOWLEDGE

Do you see any resemblance between this z-value formula and the one for the sample mean? Yes, you are right, it's basically the same. We replace $\bar{x}$ with $\bar{p}$ and then use its mean and standard deviation.

In our example, $\bar{p} = \frac{130}{300} = 0.433$, so we can calculate the z-value as follows:

$$z = \frac{0.433-0.477}{0.0288} = -1.53$$

Using the standard normal z-table in Appendix B:

$$P(\bar{p} \geq 0.433) = 1 - P(z \leq -1.53) = 1 - 0.0630 = 0.9370$$

According to this result, there is a 93.7 percent chance that 43 percent or more workers in our sample are satisfied with their jobs. Not bad at all! The shaded area in Figure 13.5 represents this probability, which displays the sampling distribution of the proportion for this example.

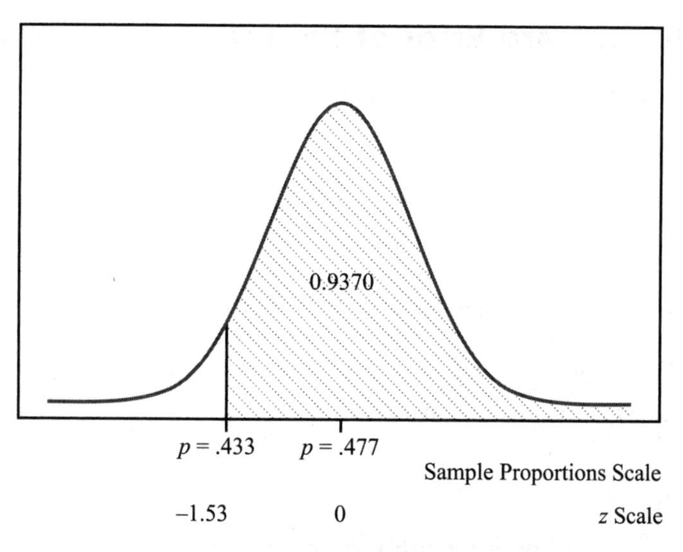

Figure 13.5
Sampling distribution of the proportion.

Practice Problems

1. Calculate the standard error of the mean when …
   a. $\sigma = 10, n = 15$
   b. $\sigma = 4.7, n = 12$
   c. $\sigma = 7, n = 20$

2. A population has a mean value of 16.0 and a standard deviation of 7.5. Calculate the following with a sample size of 9.
   a. $P(\bar{x} \leq 17)$
   b. $P(\bar{x} > 18)$
   c. $P(14.5 \leq \bar{x} \leq 16.5)$

3. Calculate the standard error of the proportion when …
   a. $p = 0.25, n = 200$
   b. $p = 0.42, n = 100$
   c. $p = 0.06, n = 175$

4. A population proportion has been estimated at 0.32. Calculate the following with a sample size of 160.

   a. $P(\bar{p} \leq 0.30)$

   b. $P(\bar{p} \leq 0.36)$

   c. $P(0.29 \leq \bar{p} \leq 0.37)$

5. A hypothetical statistics author is obsessed with making 10-foot putts. Each day that he practices, he putts 60 times and counts the number he makes. Over the last 20 practice sessions, he has averaged 24 made putts. What is the probability that he will make at least 30 putts during his next session?

The Least You Need to Know

- The sampling distribution of the sample means refers to the pattern of sample means that will occur as samples are drawn from the population at large.
- According to the central limit theorem, as the sample size, *n*, gets larger, the sample means tend to follow a normal probability distribution.
- The standard error of the mean is the standard deviation of sample means and can be determined by $\sigma_{\bar{X}} = \frac{\sigma}{\sqrt{n}}$.
- The sampling distribution of the proportion refers to the pattern of sample proportions that will occur as samples are drawn from the population.
- The standard error of the proportion is the standard deviation of the sample proportions and can be calculated by $\sigma_p = \sqrt{\frac{p(1-P)}{n}}$.

CHAPTER

14

Confidence Intervals

Now that we have learned how to collect a random sample and how sample means and sample proportions behave under certain conditions, we are ready to put those samples to work using confidence intervals.

One of the most important roles that statistics plays in today's world is to gather information from a sample and use that information to make a statement about the population from which it was chosen. We are using the sample as an estimate for the population. But just how good of an estimate is the sample providing us? The concept of confidence intervals will provide us with that answer.

In This Chapter

- Interpreting the meaning of a confidence interval
- Calculating the confidence interval for the mean when the population standard deviation is known and when it is unknown
- Introducing the Student's *t*-distribution
- Calculating the confidence interval for the proportion
- Determining sample sizes to attain a specific margin of error

The Basics of Confidence Intervals

As we have been saying over and over again, one of the most important functions of inferential statistics is to use sample information to make inferences about the population from which the sample is drawn. The confidence interval tells you how good of an estimate these inferences are. As you'll see, we will use sample information to come up with an interval that contains the population parameter of interest. But wait, there is more! We are also going to assign a confidence level to this interval. You'll be able to say that you're 90 percent or 95 percent or 99 percent confident that the interval contains the population parameter, even though you don't know what the population parameter is.

Since we are using the sample statistic (such as the sample mean) as an estimate for the population parameter, we need to distinguish between the point estimate and the interval estimate. Read on.

Point Estimate and Interval Estimate

The simplest estimate of a population is the *point estimate*, the most common being the sample mean and the sample proportion. A point estimate is a single value that best describes the population of interest. Let me explain this concept by using the following example. Let's say I'm interested in knowing the average number of carbohydrates in a bowl of cereal. So I collected a random sample of 30 different cereals and found that the average is 29 grams of carbohydrates. I could use that as my point estimate for the number of carbohydrates in the population of all cereals.

The advantage of a point estimate is that it is easy to calculate and easy to understand. The disadvantage, however, is that I have no clue as to how representative of the population it really is.

To deal with this uncertainty, we can use an *interval estimate*, which provides a range of values within which the population parameter *may* lie. Confidence intervals provide us with a confidence level that the population parameter is within the interval.

A **point estimate** is a single value that best describes the population of interest, the sample mean and sample proportion being the most common. An **interval estimate** provides a range of values that best describes the population.

The Principle of Confidence Intervals

Let's start with the confidence interval for the population mean using a large sample size, which generally refers to $n \geq 30$. The large sample enables us to use the Central Limit Theorem (yes, it comes back!). The Central Limit Theorem assures us that with our large sample, the sample mean will be normally distributed regardless of the population distribution.

To develop a confidence interval estimate, we need to learn about confidence levels. A *confidence level* is the probability that the interval estimate will include the population parameter. The common levels used by statisticians are 90 percent, 95 percent, and 99 percent. In our cereal example, let's say the 95 percent confidence interval is 25.87 to 32.13 grams of carbohydrates. (We will see in the next section how to get this interval.) This means that I'm 95 percent confident that the average number of grams of carbohydrates in all cereals is between 25.87 and 32.13.

A **confidence level** is the probability that the interval estimate will include the population parameter, such as the mean.

The Probability of an Error (the Alpha Value)

For the 95 percent confidence interval, 95 percent of all intervals will contain the population mean. The other 5 percent of the intervals won't contain the population mean. This is called the alpha (α) value, the *level of significance*, or the probability of making a Type I error. (Yes, there is Type II error.) We can write α as

$$\alpha = 1 - \text{confidence level}$$

For example, the significance level for a 90 percent confidence interval is 10 percent, the significance level for a 99 percent confidence interval is 1 percent, and so on. In general, a $(1 - \alpha)$ confidence interval has a significance level equal to α.

The **level of significance** (α) is the probability of making a Type I error.

We will revisit the level of significance in more detail in later chapters.

Confidence Intervals for the Population Mean

We will start by calculating the confidence intervals for instances when σ, the population standard deviation, is known, and then we'll move on to cases when σ is unknown.

When the Population Standard Deviation Is Known

Since the sample means follow the normal probability distribution because of the central limit theorem, we can use the standard normal distribution table to get the value for z that corresponds to our confidence level. How do we do that? Let's go back to our cereal carbohydrate example. I want to find a 95 percent confidence interval for the population mean, which is the average number of grams of carbohydrates in all cereals. So far, we know that the sample of 30 cereals has a mean of 29 grams, but I still need two more pieces of information: the population standard deviation (σ) and the z-value corresponding to the 95 percent confidence level. Let's say σ is 8.74 grams. Now how do we get the z-value?

As Figure 14.1 shows, the shaded area represents the 95 percent confidence level we chose. The un-shaded areas represent the α value, which is 0.05. Since the un-shaded area is divided into two equal areas (because the normal distribution is symmetrical around the mean), each tail represents 0.05/2 = 0.025. To get the value of z, look at the standard normal table in Appendix B and search inside the body of the table for the closest number to 0.0250. (Hint: Look at the negative z-value side because right now we are to the left of the mean.) You will find 0.0250 in the -1.9 row and 0.06 column. So our z-value to the left of the mean is -1.96. Since the distribution is symmetrical, I know you just want to use the same value, but positive, for the right side of the mean. Good idea! Your z-value is +1.96 for the right tail. Now, I can tell you understand the normal distribution pretty well!

Since we will be using these z-values a lot, it might be a good idea to keep them handy. I'll list the common ones for you:

99% confidence level, $z = \pm 2.58$

95% confidence level, $z = \pm 1.96$

90% confidence level, $z = \pm 1.65$

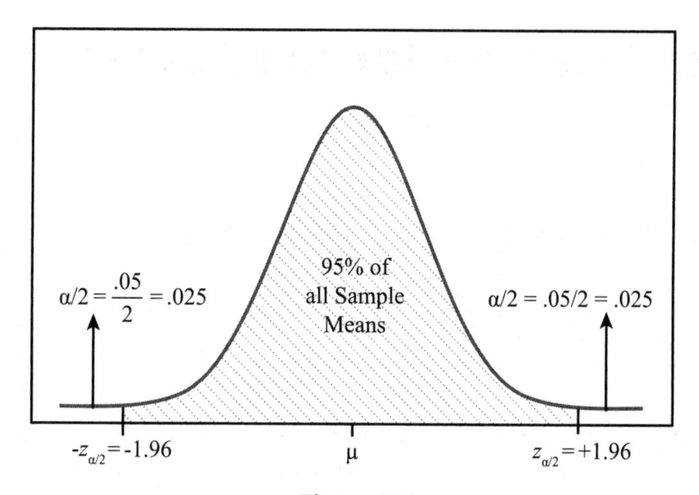

Figure 14.1
A 95 percent confidence interval.

Now we are ready to find the confidence interval. We can construct a *confidence interval* around our sample mean using the following equation:

$$\bar{x} \pm z_{\alpha/2}\sigma_{\bar{X}}$$

The upper limit of the confidence interval is:

$$\bar{x} + z_{\alpha/2}\sigma_{\bar{X}}$$

The lower limit of the confidence interval is:

$$\bar{x} - z_{\alpha/2}\sigma_{\bar{X}}$$

where:

$\bar{x}$ = the sample mean

$z_{\alpha/2}$ = the critical z-value, which is the number of standard deviations based on the confidence level

$\sigma_{\bar{X}}$ = the standard error of the mean (remember our friend from Chapter 13?)

The term $z_{\alpha/2}\sigma_{\bar{x}}$ is referred to as the *margin of error*, or *ME*, a phrase often referred to in polls and surveys.

A **confidence interval** is a range of values used to estimate a population parameter and is associated with a specific confidence level. The **margin of error (ME)** determines the width of the confidence interval and is calculated as $z_{\alpha/2}\sigma_{\bar{x}}$.

Going back to our cereal carbohydrate example, the average number of grams of carbohydrates in our sample of 30 is 29 grams, and the population standard deviation is 8.74 grams. (This represents the variation among different cereals in the population.) We can calculate our 95 percent confidence interval as follows:

$\bar{x} = 29$ grams, $n = 30$, $\sigma = 8.74$ grams, and $z_{\alpha/2} = \pm 1.96$

$\sigma_{\bar{X}} = \frac{\sigma}{\sqrt{n}} = \frac{8.74}{\sqrt{30}} = 1.5957$

Upper limit = $\bar{x} + 1.96\sigma_{\bar{X}} = 29 + 1.96(1.5957) = 32.13$

Lower limit = $\bar{x} - 1.96\sigma_{\bar{X}} = 29 - 1.96(1.5957) = 25.87$

According to this result, our 95 percent confidence interval for this random sample of cereal carbohydrates is between 25.87 and 32.13 grams. In other words, we are 95 percent confident that the number of grams of carbohydrates for all cereals is between 25.87 and 32.13 (or 25.87, 32.13).

Another way to get this interval is by using the margin of error. The margin of error $(ME) = z_{\alpha/2}\sigma_{\bar{x}}$. So in our example, $ME = (1.96)(1.5957) = 3.13$. The 95 percent confidence interval is $\bar{x} \pm ME = 29 \pm 3.13 = (25.87, 32.13)$. As you can see from this example, finding the 95 percent confidence interval this way is very simple. We start with the sample mean, then we add the margin of error to get the upper limit of the interval and subtract the margin of error to get the lower limit of the interval. Easy, isn't it? This interval is shown in Figure 14.2.

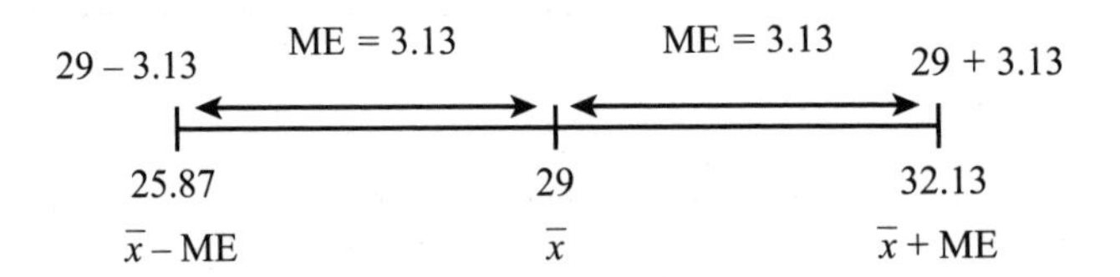

Figure 14.2
A 95 percent confidence interval for cereal's average number of grams of carbohydrates.

Beware of Interpretation of the Confidence Interval!

As described previously, a confidence interval is a range of values used to estimate a population parameter and is associated with a specific confidence level. A confidence interval needs to be described in the context of several samples. If we select 20 different cereal samples from our population and construct 95 percent confidence intervals around each of the sample means, then theoretically 19 of the 20 intervals (95 percent of all samples) will contain the true population mean, which remains unknown. Figure 14.3 shows this concept. For all samples with an average ($\bar{x}$) in the shaded area, the confidence intervals for these sample means will contain the true population mean. Only for 5 percent of all samples—those with a very small average (left tail) or a very large average (right tail)—will the confidence intervals not contain the true population mean. Figure 14.3 shows some of the 95 percent confidence intervals. As you can see, all of them except one, $\bar{x}_5$, do contain the true population mean.

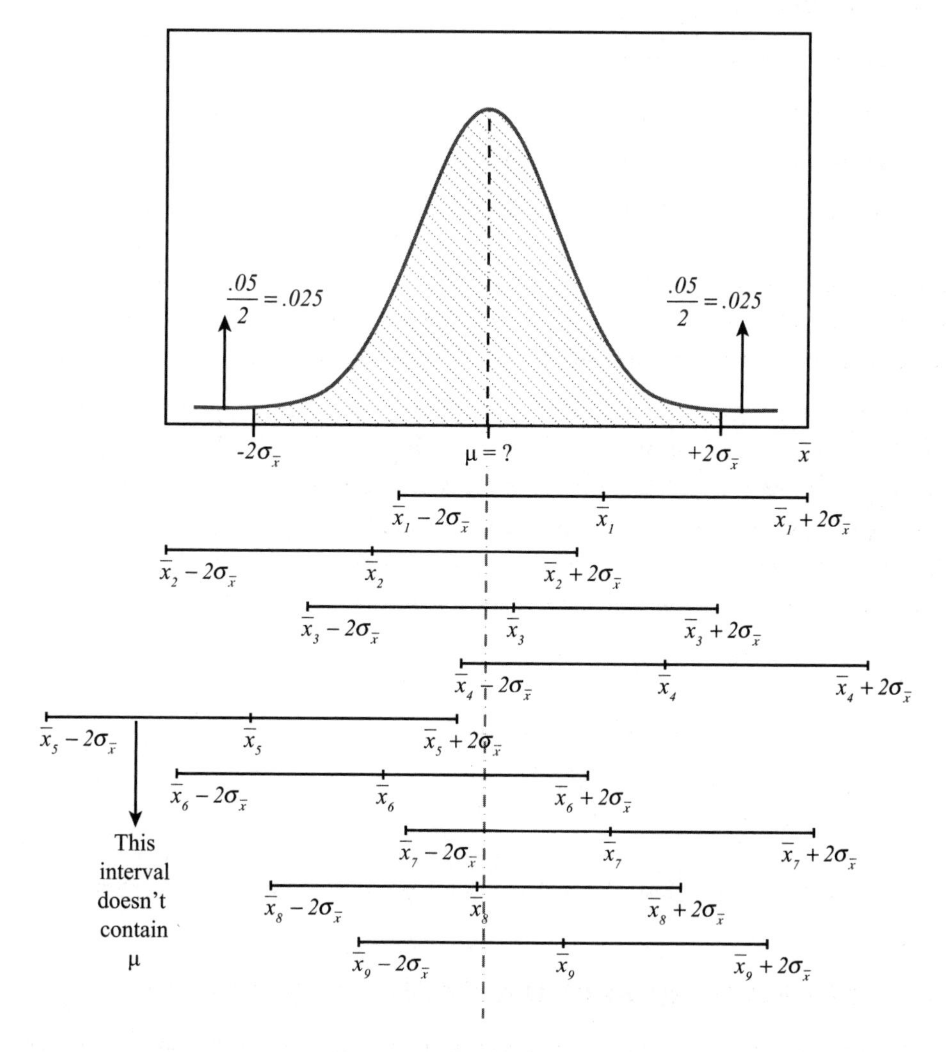

Figure 14.3
Interpreting the definition of a confidence interval.

WRONG NUMBER

It is easy to misinterpret the definition of a confidence interval. For example, it is *not* correct to state "there is a 95 percent probability that the true population mean is within the interval 25.87 and 32.13 grams of carbohydrates." Rather, a correct statement would be "there is a 95 percent probability that any given confidence interval from a random sample will contain the true population mean."

Because there is a 95 percent probability that any given confidence interval will contain the true population mean in the previous example, we have a 5 percent chance that it won't. This 5 percent value is the *level of significance*, α, which is represented by the total white area in both tails of Figure 14.3.

The Effect of Changing Confidence Levels

So far, we have only referred to a 95 percent confidence interval. However, we can choose other confidence levels to suit our needs. The following table shows our cereal carbohydrate example with confidence levels of 90, 95, and 99 percent.

Confidence Intervals with Various Confidence Levels

Confidence Level	$z_{\alpha/2}$	$\sigma_{\bar{X}}$	Sample Mean	*ME*	Lower Limit	Upper Limit
90	1.65	1.5957	29	2.63	26.37	31.63
95	1.96	1.5957	29	3.13	25.87	32.13
99	2.58	1.5957	29	4.12	24.88	33.12

From the previous table, you can see that there's a price to pay for increasing the confidence level—our interval estimate of the true population mean becomes wider. We have proven that, once again, there is no free lunch with statistics. If you want more certainty that your confidence interval will contain the true population mean, then your confidence interval will become wider.

I've a feeling that you notice something else: as the confidence level increases, so does the margin of error. Yes, that's why the interval gets wider.

The Effect of Changing the Sample Size

There is one way, however, to reduce the width of our confidence interval while maintaining the same confidence level. We can do this by increasing the sample size. There is still no free lunch though because increasing the sample size has a cost associated with it. Let's say we increase our sample size to include 62 different cereals. This change will affect our standard error as follows:

$$\sigma_{\bar{X}} = \frac{\sigma}{\sqrt{n}} = \frac{8.74}{\sqrt{62}} = 1.11 \text{ grams}$$

Our new 95 percent confidence interval for our original sample will be:

$\bar{x} = 29$ grams, $n = 62$, $\sigma_{\bar{X}} = 1.11$ grams

Upper limit = $\bar{x} + 1.96\sigma_{\bar{X}} = 29 + 1.96(1.11) = 31.18$ grams

Lower limit = $\bar{x} - 1.96\sigma_{\bar{X}} = 29 - 1.96(1.11) = 26.82$ grams

Increasing our sample size from 30 to 62 has reduced the 95 percent confidence interval from (25.87, 32.13) to (26.82, 31.18). Can you tell what will happen to the margin of error when I increase my sample size? It will also decrease. As n increases to 62, the margin of error is now $z_{\alpha/2}\sigma_{\bar{x}} = 1.96(1.11) = 2.18$ grams, compared with 3.13 grams when $n = 30$.

So far we have seen how to get the confidence interval for the population mean when sigma is known and we have a large sample ($n \geq 30$). What if we have a small sample instead? The bad news, in this case, is we can't use the central limit theorem, so the population from which the sample is drawn *must* be normally distributed. The good news is, if the population is normally distributed, then we can construct the confidence interval for a small sample the same way as we did with the large sample.

When the Population Standard Deviation Is Unknown

Here's a simple section for you. (It's about time!) So far we have assumed that we know σ, the population standard deviation. What happens if σ is unknown? Don't panic, we can substitute s, the sample standard deviation, for σ, the population standard deviation, and then follow the same procedure as before as long as we have a large sample. In this case, the confidence interval will be $\bar{x} \pm z_{\alpha/2}s_{\bar{x}}$ where $s_{\bar{x}} = \frac{s}{\sqrt{n}}$. Let's see an example to illustrate this large sample case, and then we'll move to the small sample instance.

Consider the following table that shows the number of grams of carbohydrates in a sample of 32 different cereals.

Cereal Carbohydrate Grams Sample ($n = 32$)

32	23	32	47	30	22	26	43
29	36	33	22	45	25	29	46
24	22	27	41	42	25	20	28
25	13	44	24	25	37	18	35

Using Excel, we can confirm that:

$\bar{x} = 30$ grams and $s = 8.97$ grams

A 99 percent confidence interval around this sample mean would be:

$$\bar{x} \pm 2.58 s_{\bar{x}}$$

$$s_{\bar{x}} = \frac{s}{\sqrt{n}} = \frac{8.97}{\sqrt{32}} = 1.59$$

$$\text{Upper limit} = \bar{x} + 2.58 s_{\bar{x}} = 30 + 2.58(1.59) = 34.1 \text{ grams}$$

$$\text{Lower limit} = \bar{x} - 2.58 s_{\bar{x}} = 30 - 2.58(1.59) = 25.9 \text{ grams}$$

We can also use the margin of error approach here. In this example, $ME = 2.58(1.59) = 4.10$. When you subtract this from 30, you get the lower limit as 25.9 grams, and when you add it to 30, you get the upper limit of the interval as 34.1 grams.

See! That wasn't too bad.

When the Population Standard Deviation Is Unknown and with Small Samples

When σ is unknown and we have a small sample, we can't use the standard normal distribution anymore. This substitution forces us to use a new probability distribution known as the Student's *t*-distribution (named in honor of you, the student).

RANDOM THOUGHTS

The Student's *t*-distribution was developed by William Gosset (1876–1937) while working for the Guinness Brewing Company in Ireland. He published his findings using the pseudonym Student. Now there's a rare statistical event—a bashful Irishman!

The *t*-distribution is a continuous probability distribution with the following properties:

- It is bell-shaped and symmetrical around the mean.
- The shape of the curve depends on the *degrees of freedom* (*d.f.*) which, when dealing with the sample mean, would be equal to $n - 1$.
- The area under the curve is equal to 1.0.
- The *t*-distribution has a mean of zero, just like the normal distribution, but a variance greater than one.

- The *t*-distribution is flatter than the normal distribution. As the number of degrees of freedom increases, the shape of the *t*-distribution becomes similar to the normal distribution as seen in Figure 14.4. With more than 30 degrees of freedom (a sample size of 30 or more), the two distributions are practically identical.

The **degrees of freedom** are the number of values that are free to be varied given information, such as the sample mean, is known.

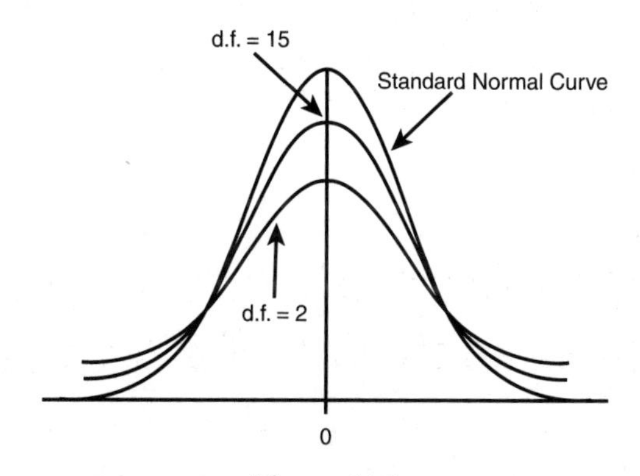

Figure 14.4
The Student's t-distribution compared to the normal distribution.

Students often struggle with the concept of degrees of freedom, which represent the number of remaining free choices you have after something has been decided, such as the sample mean. For example, if I know that my sample of size 3 has a mean of 10, I can only vary two values ($n - 1$). After I set those two values, I have no control over the third value because my sample average must be 10. For this sample, I have 2 degrees of freedom.

We can now set up our confidence intervals for the mean using a small sample: $\bar{x} \pm t_{\alpha/2} s_{\bar{x}}$

The upper limit of the confidence interval is $\bar{x} + t_{\alpha/2} s_{\bar{x}}$

The lower limit of the confidence interval is $\bar{x} - t_{\alpha/2} s_{\bar{x}}$

where:

$t_{\alpha/2}$ = critical t-value (can be found in Table 4 in Appendix B)

$s_{\bar{x}} = \frac{s}{\sqrt{n}}$ = the estimated standard error of the mean

To demonstrate this procedure, let's assume the population of cereals' grams of carbohydrates follows a normal distribution and the following sample of 10 cereals were collected.

Number of Grams of Carbohydrates in a Cereal Sample from a Normal Distribution ($n = 10$)

36	22	13	32	15	27	44	18	22	41

With σ unknown, we will construct a 95 percent confidence interval around the sample mean as follows:

To determine the value of $t_{\alpha/2}$ for this example, I need to calculate the number of degrees of freedom. Because $n = 10$, I have $n - 1 = 9$ *d.f.* This corresponds to $t_{\alpha/2} = 2.262$, which is underlined in the following table taken from Table 4 in Appendix B.

Excerpt from the Student's *t*-Distribution:

Selected right-tail areas with confidence levels underneath

Alpha	**0.2000**	**0.1500**	**0.1000**	**0.0500**	**0.0250**	**0.0100**	**0.0050**	**0.0010**	**0.0005**
Conf lev	**0.6000**	**0.7000**	**0.8000**	**0.9000**	**0.9500**	**0.9800**	**0.9900**	**0.9980**	**0.9990**
d.f.									
1	1.376	1.963	3.078	6.314	12.706	31.821	63.657	318.31	636.62
2	1.061	1.386	1.886	2.920	4.303	6.965	9.925	22.327	31.599
3	0.978	1.250	1.638	2.353	3.182	4.541	5.841	10.215	12.924
4	0.941	1.190	1.533	2.132	2.776	3.747	4.604	7.173	8.610
5	0.920	1.156	1.476	2.015	2.571	3.365	4.032	5.893	6.869
6	0.906	1.134	1.440	1.943	2.447	3.143	3.707	5.208	5.959
7	0.896	1.119	1.415	1.895	2.365	2.998	3.499	4.785	5.408
8	0.889	1.108	1.397	1.860	2.306	2.896	3.355	4.501	5.041
9	0.883	1.100	1.383	1.833	2.262	2.821	3.250	4.297	4.781
10	0.879	1.093	1.372	1.812	2.228	2.764	3.169	4.144	4.587

We next need to calculate the sample mean and sample standard deviation, which, according to Excel, are as follows:

$$\bar{x} = 27 \text{ grams and } s = 10.83 \text{ grams}$$

We can calculate the standard error of the mean:

$$s_{\bar{x}} = \frac{s}{\sqrt{n}} = \frac{10.83}{\sqrt{10}} = 3.42$$

and can construct our 95 percent confidence interval:

$$\text{Upper limit} = \bar{x} + t_{\alpha/2} s_{\bar{x}} = 27 + 2.262(3.42) = 34.74 \text{ grams}$$

$$\text{Lower limit} = \bar{x} - t_{\alpha/2} s_{\bar{x}} = 27 - 2.262(3.42) = 19.26 \text{ grams}$$

We can still use the margin of error approach as before. In this example, the margin of error is 7.74 grams. I'll leave it to you to confirm the confidence interval.

Now that wasn't too bad!

BOB'S BASICS

We can use the t-distribution when all of the following conditions have been met:

- The population follows the normal (or approximately normal) distribution.
- The sample size is less than 30.
- The population standard deviation, σ, is unknown and must be approximated by s, the sample standard deviation.

Determining Sample Size for the Population Mean

Knowing the appropriate sample size needed for a specific confidence level and margin of error is very important, especially for quality control. For example, M&M's bags weigh 1.69 ounces. The producers of M&M's want to make sure that the bags weigh 1.69 ounces. Weighing every single M&M's bag is costly and time consuming, so M&M's producers want to use a sample of M&M's, weigh them, and check if the weight is 1.69 ounces or not. How large should the sample be? This is what this section is all about.

We can calculate a minimum sample size that would be needed to provide a specific margin of error. In our cereal carbohydrates example, what sample size would we need for a 95 percent confidence interval that has a margin of error of ±3 grams?

$$ME = z_{\alpha/2}\sigma_{\bar{x}}$$

$$ME = \frac{z_{\alpha/2}\sigma}{\sqrt{n}}$$

$$\sqrt{n} = \frac{z_{\alpha/2}\sigma}{ME}$$

$$n = \left(\frac{z_{\alpha/2}\sigma}{ME}\right)^2 = \frac{(z_{\alpha/2})^2\sigma^2}{(ME)^2}$$

$$n = \frac{(1.96)^2(8.74)^2}{(3)^2} = 32.61 \approx 33$$

Therefore, to obtain a 95 percent confidence interval that ranges from 29 – 3 = 26 grams to 29 + 3 = 32 grams would require a sample size of 33 cereals.

If you want to reduce your margin of error to ±2 grams, you will need a larger sample of ≈74 cereals! Remember, there is no free lunch in statistics!

Using Excel's CONFIDENCE Function

Excel has a pretty cool built-in function that calculates confidence intervals for us. If you are using the normal distribution, then use the CONFIDENCE.NORM function, and if you are using the *t*-distribution, then use the CONFIDENCE.T function. The CONFIDENCE.NORM function has the following characteristics:

CONFIDENCE.NORM(alpha, standard_dev, size)

where:

alpha = the significance level of the confidence interval

standard_dev = the standard deviation of the population

size = sample size

For instance, Figure 14.5 shows the CONFIDENCE.NORM function being used to calculate the confidence interval for our original cereal carbohydrates example.

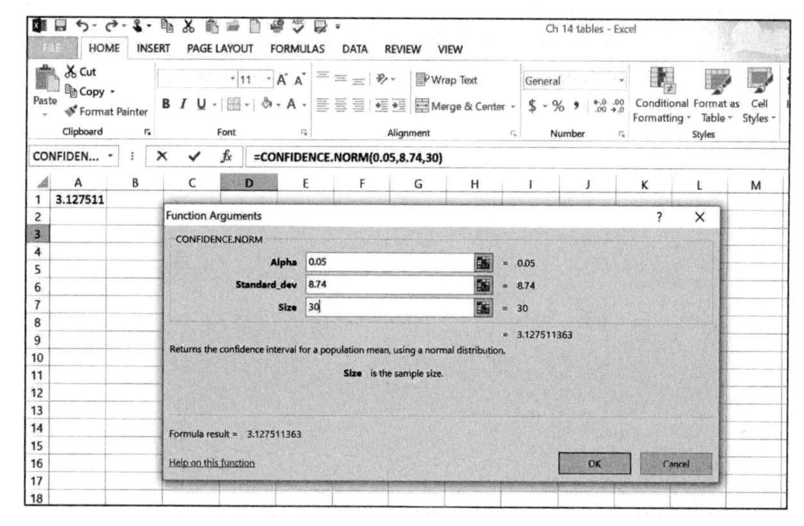

Figure 14.5

CONFIDENCE.NORM function in Excel for the cereal carbohydrates example.

Cell A1 contains the Excel formula =CONFIDENCE.NORM(0.05,8.74,30) with the result being 3.127511. This value represents the margin of error, or the amount to add and subtract from the sample mean, as follows:

$$29 + 3.13 = 32.13 \text{ grams}$$

$$29 - 3.13 = 25.87 \text{ grams}$$

This confidence interval is slightly different from the one calculated earlier in the chapter due to the rounding of numbers.

Now, let's use Excel to calculate the confidence interval using the *t*-distribution. The CONFIDENCE.T function has the following characteristics:

CONFIDENCE.T(alpha, standard_dev, size)

where:

alpha = the significance level of the confidence interval

standard_dev = the standard deviation of the sample

size = sample size

For instance, Figure 14.6 shows the CONFIDENCE.T function being used to calculate the confidence interval for our cereal carbohydrates example with a small sample and σ unknown.

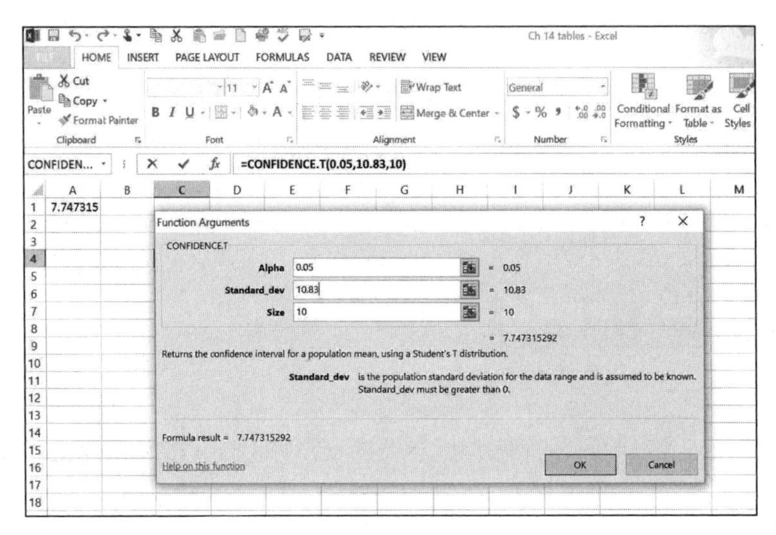

Figure 14.6

CONFIDENCE.T function in Excel for the cereal carbohydrates example.

Cell A1 contains the Excel formula =CONFIDENCE.T(0.05,10.83,10) with the result being 7.747315. As with CONFIDENCE.NORM, this value represents the margin of error, or the amount to add and subtract from the sample mean, as follows:

$$27 + 7.75 = 34.75 \text{ grams}$$

$$27 - 7.75 = 19.25 \text{ grams}$$

This confidence interval is also slightly different from the one calculated earlier in the chapter due to the rounding of numbers. This sure beats using tables and square root functions on the calculator.

That ends our discussion on confidence intervals around the mean. Next on the menu is proportions!

Confidence Intervals for the Population Proportion with Large Samples

We can also estimate the proportion of a population by constructing a confidence interval from a sample. As you might recall from Chapter 13, proportion data follow the binomial distribution that can be approximated by the normal distribution under the following conditions:

$$np \geq 5 \text{ and } nq \geq 5$$

where:

p = the probability of a success in the population

q = the probability of a failure in the population ($q = 1 - p$)

Suppose I want to estimate the proportion of high school graduates who enroll in colleges or universities based on the results of a sample. In Chapter 13, we learned that we can calculate the proportion of a sample using:

$$\bar{p} = \frac{\text{Number of Successes in the Sample}}{n}$$

Calculating the Confidence Interval for the Population Proportion

To construct the confidence interval around the sample proportion, we need to know the standard error of the proportion. As we saw in Chapter 13, the standard error of the proportion (σ_p) can be calculated by:

$$\sigma_p = \sqrt{\frac{p(1-P)}{n}}$$

There's extra credit for anyone who can see a problem arising here. Our challenge is that we are trying to estimate p, the population proportion, but we need a value for p to set up the confidence interval. Our solution—estimate the standard error by using the sample proportion, $\bar{p}$, as an approximation for the population proportion as follows:

$$\hat{\sigma}_p = \sqrt{\frac{\bar{p}(1-\bar{P})}{n}}$$

We now can construct a confidence interval around the sample proportion by:

$\bar{p} + z_{\alpha/2}\hat{\sigma}_p$ (upper limit of the confidence interval)

$\bar{p} - z_{\alpha/2}\hat{\sigma}_p$ (lower limit of the confidence interval)

Let's put these equations to work. In my efforts to estimate the proportion of high school graduates who enroll in colleges or universities, I sample 175 random high school graduates, of whom 110 enrolled in colleges or universities. I can now calculate $\bar{p}$, the sample proportion:

$$\bar{p} = \frac{\text{Number of Successes in the Sample}}{n} = \frac{110}{175} = 0.629$$

The estimated standard error of the proportion would be:

$$\hat{\sigma}_p = \sqrt{\frac{\bar{p}(1-\bar{p})}{n}} = \sqrt{\frac{(0.629)(1-0.629)}{175}} = 0.0365$$

We are now ready to construct a 95 percent confidence interval around our sample proportion ($z_{\alpha/2} = 1.96$):

$$\text{Upper limit} = \bar{p} + 1.96\,\hat{\sigma}_p = 0.629 + 1.96(0.0365) = 0.70$$

$$\text{Lower limit} = \bar{p} - 1.96\,\hat{\sigma}_p = 0.629 - 1.96(0.0365) = 0.557$$

Our 95 percent confidence interval for the proportion of high school graduates who enroll in colleges or universities is (0.56, 0.70), so we are 95 percent confident that between 56 percent and 70 percent of high school graduates enroll in colleges and universities. A reality check, according to the Bureau of Labor Statistics: 68.4 percent of high school graduates enrolled in colleges and universities in 2014. That's well within our interval!

You can also use the margin of error approach to get the confidence interval for the proportion. In our example, the margin of error is 1.96(0.0365) = 0.0715. Subtracting it and adding it to the mean yields the same confidence interval as before. This is especially useful for the polls we hear in the news. For example, according to Gallup, President Obama's approval rating on November 5-7, 2015 was 48 percent with a margin of error of ±3 percentage points. This means that the 95 percent confidence interval for President Obama's approval rating is 45 percent to 51 percent. See, now you understand the small print better!

Determining Sample Size for the Proportion

Just as we did for the mean, we can determine a required sample size that would be needed to provide a specific margin of error. What sample size would we need for a 99 percent confidence interval that has a margin of error of ±6 percent ($ME = 0.06$) in our example of high school graduates enrolling in college and universities? The formula to calculate n, the sample size is:

$$n = \frac{(z_{\alpha/2})^2\,\bar{p}(1-\bar{p})}{(ME)^2}$$

Notice that we need a value for $\bar{p}$ here. We have a couple of options:

- If we have a preliminary estimate of $\bar{p}$ from a previous study, we can use it.
- If we don't have a preliminary estimate, then set $p = q = 0.50$.

Applying this to our example,

$$n = \frac{(2.58)^2(0.5)(1-0.5)}{(0.06)^2} = 462.25 \approx 463$$

Therefore, to obtain a 99 percent confidence interval that provides a margin of error no more than ±6 percent would require a sample size of 463 high school graduates.

RANDOM THOUGHTS

The reason we use $p = q = 0.50$ if we don't have an estimate of the population proportion is that these values provide the largest sample size when compared to other combinations of p and q. It's like being penalized for not having specific information about your population. This way you are sure your sample size is large enough, regardless of the population proportion.

Practice Problems

1. Construct a 97 percent confidence interval around a sample mean of 31.3 taken from a population that is not normally distributed with a standard deviation of 7.6 using a sample of size 40.
2. What sample size would be necessary to ensure a margin of error of 5 for a 98 percent confidence interval taken from a population that is not normally distributed and which has a population standard deviation of 15?
3. Construct a 90 percent confidence interval around a sample mean of 16.3 taken from a population that is not normally distributed with a population standard deviation of 1.8 using a sample of size 10.
4. The following sample of size 30 was taken from a population that is not normally distributed:

10	4	9	12	5	17	20	9	4	15
11	12	16	22	10	25	21	14	9	8
14	16	20	18	8	10	28	19	16	15

   Construct a 90 percent confidence interval around the mean.

5. The following sample of size 12 was taken from a population that is normally distributed and that has a population standard deviation of 12.7:

37	48	30	55	50	46	40	62	50	43	36	66

   Construct a 94 percent confidence interval around the mean.

6. The following sample of size 11 was taken from a population that is normally distributed:

121	136	102	115	126	106	115	132	125	108	130

   Construct a 98 percent confidence interval around the mean.

7. The following sample of size 11 was taken from a population that is not normally distributed:

   87 59 77 65 98 90 84 56 75 96 66

   Construct a 99 percent confidence interval around the mean.

8. A sample of 200 light bulbs was tested, and it was found that 11 were defective. Calculate a 95 percent confidence interval around this sample proportion.

9. What sample size would you need to construct a 96 percent confidence interval around the proportion for voter turnout during the next election that would provide a margin of error of 4 percent? Assume the population proportion has been estimated at 55 percent.

The Least You Need to Know

- A confidence interval is a range of values used to estimate a population parameter and is associated with a specific confidence level.
- A confidence level is the probability that the interval estimate will include the population parameter, such as the mean.
- Increasing the confidence level results in the confidence interval becoming wider.
- Increasing the sample size reduces the width of the confidence interval.
- Use the *t*-distribution to construct a confidence interval when the population follows the normal (or approximately normal) distribution, the sample size is less than 30, and the population standard deviation, σ, is unknown.
- Use the normal distribution to construct a confidence interval around the sample proportion when $np \geq 5$ and $nq \geq 5$.

CHAPTER

15

Hypothesis Testing with One Population

Now that we know how to make an estimate of a population parameter, such as a mean, using a sample and a confidence interval, let's move on to the heart and soul of inferential statistics: hypothesis testing.

One thing statisticians like to do is to make a statement about a population parameter, collect a sample from that population, measure the sample, and declare, in a scholarly manner, whether or not the sample supports the original statement. This, in a nutshell, is what hypothesis testing is all about. Of course, I've included a few juicy details. Without them, this would be one short chapter!

In this chapter, we will introduce you to the concept of hypothesis testing and apply it to hypothesis testing that involves only one population.

Hypothesis testing involving one population focuses on confirming claims such as the population average is equal to a specific value. We will consider many different cases with this type of hypothesis test in the following sections. This chapter relies on some of the concepts we explored in Chapter 14, so be sure you are comfortable with that material before you dive into this chapter.

In This Chapter

- Formulating the null and alternative hypotheses
- Distinguishing between a one-tail and two-tail hypothesis test
- Testing the mean of a population when the population standard deviation is known and when it is unknown
- Examining the role of alpha (α) in hypothesis testing
- Using the p-value to test a hypothesis
- Testing the proportion of a population using a large sample

Hypothesis Testing: The Traditional Method

In the statistical world, a *hypothesis* is an assumption about a population parameter, such as the population mean or the population proportion. Examples of hypotheses (that's plural for hypothesis) include the following:

- The average adult drinks 3.1 cups of coffee per day.
- The average student debt for a college graduate for the class of 2015 was $35,051 (the highest in U.S. history).
- According to the Department of Transportation, the average age of passenger cars on the road in the United States in 2014 was 11.4 years.

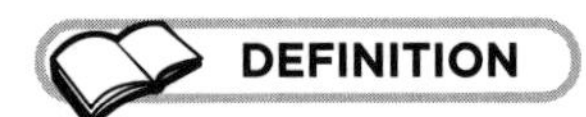

A **hypothesis** is an assumption about a population parameter that is developed for the purpose of testing.

In each case, we have made a statement about the population that may or may not be true. The purpose of hypothesis testing is to use sample information and make a statistical conclusion about rejecting or not rejecting such statements. Let's see how to do that!

Procedures for Hypothesis Testing

A study by Sallie Mae released in April 2009 showed that the average credit card debt for graduating college seniors in 2008 was $4,100. The president of our college believes that our graduating seniors don't owe $4,100 on their credit cards so he asked me to check this for him. How would I go about doing this? I use a sample of our students and get their average credit card debt. Suppose that the sample average is $3,900. Hypothesis testing will then tell me whether or not $3,900 is significantly different from $4,100, or if the difference is merely due to chance. So how do we do hypothesis testing?

There are five steps that we need to complete for hypothesis testing:

1. State the null and alternative hypotheses.
2. Determine the level of significance.
3. Calculate the test statistic.
4. Determine the critical value(s).
5. State your decision or finding.

So let's look at each one of these in detail.

1. The Null and Alternative Hypotheses

Every hypothesis test has both a null hypothesis and an alternative hypothesis. The *null hypothesis*, denoted by H_0, represents the status quo and involves stating the belief that the mean of the population is ≤, =, or ≥ a specific value. The null hypothesis is believed to be true unless there is overwhelming evidence to the contrary. The null hypothesis is the one to be rejected or not rejected. In our example, the null hypothesis would be stated as:

$$H_0{:}\mu = \$4{,}100$$

The *alternative hypothesis*, denoted by H_1, represents the opposite of the null hypothesis and holds true if the null hypothesis is found to be false. The alternative hypothesis always states the mean of the population is <, ≠, or > a specific value. In our example, the alternative hypothesis would be stated as:

$$H_0 : \mu \neq \$4{,}100$$

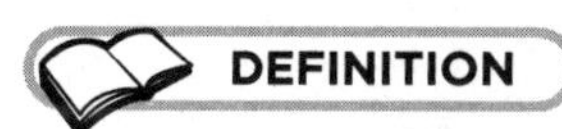

The **null hypothesis**, denoted by H_0, is a statement about the value of the population mean (μ) that needs to be tested, and takes the form μ ≤, =, or ≥ a specific value. The **alternative hypothesis**, denoted by H_1, represents the opposite of the null hypothesis and holds true if the null hypothesis is to be rejected.

The following table shows the three combinations of the null and alternative hypotheses.

Null Hypothesis	Alternative Hypothesis
$H_0 : \mu = \$4{,}100$	$H_1 : \mu \neq \$4{,}100$
$H_0 : \mu \geq \$4{,}100$	$H_1 : \mu < \$4{,}100$
$H_0 : \mu \leq \$4{,}100$	$H_1 : \mu > \$4{,}100$

When do you use each hypothesis? Good question. This depends on whether you are doing a two-tail or a one-tail test as explained in a following section.

RANDOM THOUGHTS

Some textbooks use the convention that the null hypothesis will always be stated as = and will not use ≤ or ≥. Choosing either method of stating your hypothesis will not affect the statistical analysis. Just be consistent with the convention you decide to use.

2. The Level of Significance

Remember that the purpose of hypothesis testing is to verify the validity of a claim about a population based on a single sample. Because we are relying on a sample, we expose ourselves to the risk that our conclusions about the population could be wrong because of sampling error.

In our college seniors' credit card debt example, suppose that we reject H_0. That is, according to the sample, the average credit card debt for our college seniors isn't $4,100. But what if the true population mean actually is $4,100? In other words, what if our seniors have $4,100 credit card debt but the sample we chose is not fully representative of the population? This can occur primarily because of sampling error, which we discussed in Chapter 12. This type of error, rejecting H_0 when it's actually true, is known as a *Type I error.* The probability of making a Type I error is known as α, the level of significance, which we introduced in Chapter 14.

We can also have another type of error with hypothesis testing. Let's say in our seniors' credit card debt example, we don't reject H_0. That is, according to the sample, our college seniors' credit card debt is $4,100. But what if the true population mean is actually not $4,100? This type of error, when we do not reject H_0 when in reality it's false, is known as a *Type II error.* The probability of making a Type II error is known as β.

A **Type I error** occurs when we reject the null hypothesis while in reality it is true. A **Type II error** occurs when we fail to reject the null hypothesis when in reality it is not true.

The following table summarizes the two types of hypothesis errors.

	H_0 Is True	H_0 Is False
Reject H_0	Type I Error P(Type I Error)= α	Correct Decision
Do Not Reject H_0	Correct Decision	Type II Error P(Type II Error) = β

Normally, with hypothesis testing, we decide on a value for a that is somewhere between 0.01 and 0.10 before we collect the sample. The value of β can then be calculated, but that topic goes beyond the scope of this book. Be grateful for this because that concept is very complicated!

RANDOM THOUGHTS

Ideally, we would like the values of α and β to be as small as possible. However, for a given sample size, reducing the value of α will result in an increase in the value of β. The opposite also holds true. The only way to reduce both α and β simultaneously is to increase the sample size. Once the sample size has been increased to the size of the population, the values of α and β will be 0. However, as we discussed in Chapter 12, this is not a recommended strategy.

3. The Test Statistic

Using the information from the sample, we are going to calculate a test statistic that we will use to determine whether to reject or not reject H_0. To calculate the z-test statistic, we use the following formula:

$$z = \frac{\bar{x}-\mu}{\sigma_{\bar{x}}} = \frac{\bar{x}-\mu}{\sigma/\sqrt{n}}$$

As we will see later in the chapter, we use the z-test statistic formula if σ is known. If σ is not known, we use the t-test statistic.

4. The Critical Value

The critical value divides the area under the normal distribution curve into two regions: the area where we don't reject H_0 and the area(s) where we reject H_0. We get the critical value from the table, and it differs according to whether we are doing a two-tail or a one-tail test.

I know you might be asking how I get this critical value. It's very simple! Let me explain.

Two-tail test:

If we choose $\alpha = 0.05$ and because this is a two-tail test, this area needs to be evenly divided between both tails, with each tail receiving $\alpha/2$ ($0.05/2 = 0.025$). According to Figure 15.1, we need to find the critical z-value that corresponds to the area 0.025. Using Table 3 in Appendix B, we look inside the body of the table for the closest value to 0.025. We can find this value by looking across row -1.9 and down column 0.06 to arrive at a critical value of -1.96 for the left tail. How about the right tail? Yes, you don't need to look it up—it is +1.96 since the normal distribution curve is symmetrical.

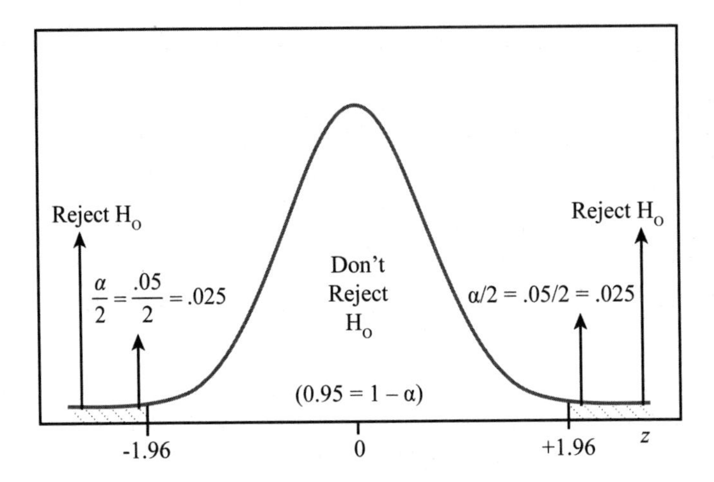

Figure 15.1
Critical value for z for a two-tail test for $\alpha = 0.05$.

One-tail test:

If we have a one-tail test, we will have only one area of rejection, not two. If it is a right tail test, then the area of rejection will be in the right tail, and if we have a left tail test, the area of rejection will be in the left tail. Let's see how to get the critical value for each one.

If we choose $\alpha = 0.01$ and use a right tail test, then we will need to determine the critical z-value that corresponds. Because this is a one-tail test, this entire area needs to be in one rejection region in the right side of the distribution. As Figure 15.2 shows, we need to find the z-value that corresponds to the area 0.99 or $1 - \alpha$.

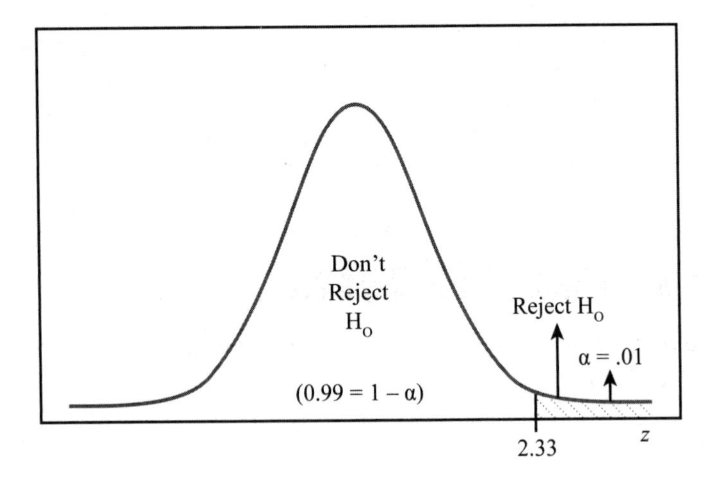

Figure 15.2
Critical value for z for a one-tail test for $\alpha = 0.01$.

Using Table 3 in Appendix B, we look inside the body of the table for the closest value to 0.9900, which results in a critical *z*-value of 2.33.

For the left tail test, the rejection area will be to the left, and the critical value will be -2.33 instead of +2.33.

5. State Your Decision

Now compare the calculated *z*-test (or *t*-test) statistic to the critical value and make your decision as follows:

- If the calculated *z*-test statistic falls within the white region, we do not reject H_0. That is, we do not have enough evidence to support H_1, the alternative hypothesis, which states that the population mean is not equal to $4,100.
- If the calculated *z*-test statistic falls in either shaded region, otherwise known as the rejection region, we reject H_0. That is, we have enough evidence to support H_1, which results in our belief that the true population mean is not equal to $4,100.

WRONG NUMBER

The only two statements that we can make about the null hypothesis are that we ...

- Reject the null hypothesis.
- Do not reject the null hypothesis.

Because our conclusions are based on a sample, we will never have enough evidence to accept the null hypothesis. It's a much safer statement to say that we do not have enough evidence to reject H_0. We can use the analogy of the legal system to explain. The jury's decision is that the defendant is "guilty" or "not guilty." If a jury finds a defendant "not guilty," they are not saying the defendant is innocent. Rather, they are saying that there is not enough evidence to prove guilt.

One-Tail vs. Two-Tail Tests

When testing our hypotheses, we can perform either a one-tail test or a two-tail test. Let's see what each one is and when to use each.

Two-Tail Hypothesis Testing

A *two-tail hypothesis test* is used whenever the alternative hypothesis is expressed as ≠. As in our seniors' credit card debt example, the null and alternative hypotheses are stated as:

$$H_0: \mu = \$4{,}100$$

$$H_1: \mu \neq \$4{,}100$$

For a two-tail test, we have two areas of rejection: one in the right tail and one in the left tail as in Figure 15.3. Since the *z*-critical values are used a lot, I'm going to list them for you so they're handy whenever you need them:

$$a = 0.01 \Rightarrow z \text{ critical value} = \pm 2.58$$

$$a = 0.05 \Rightarrow z \text{ critical value} = \pm 1.96$$

$$a = 0.10 \Rightarrow z \text{ critical value} = \pm 1.65$$

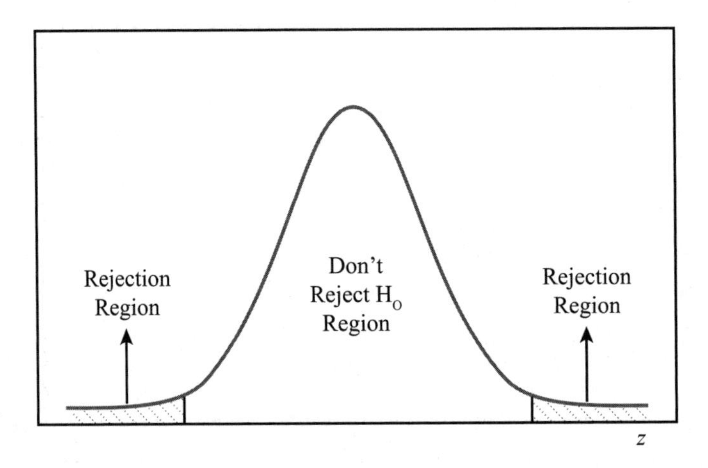

Figure 15.3
Two-tail hypothesis test.

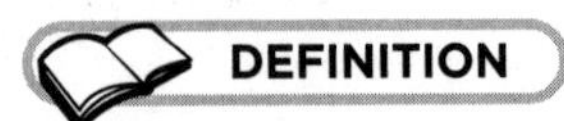

The **two-tail hypothesis test** is used whenever the alternative hypothesis is expressed as ≠.

Because there are two rejection regions in this figure, we have a two-tail hypothesis test. If the calculated test statistic falls in either tail (the rejection regions), then we reject the null hypothesis.

One-Tail Hypothesis Testing

A *one-tail hypothesis test* involves the alternative hypothesis being stated as < or >. Let's say that our president believes that our seniors have less credit card debt than the national average of $4,100. To test this, we'll use a left tail test. The null and alternative hypotheses are stated as:

$$H_0 : \mu \geq \$4{,}100$$

$$H_1 : \mu < \$4{,}100$$

For a one-tail test, we have only one area of rejection. In this case, it is in the left tail of the distribution, which is the shaded area in Figure 15.4. The critical z-values in this case are:

$$a = 0.01 \Rightarrow z \text{ critical value} = -2.33$$

$$a = 0.05 \Rightarrow z \text{ critical value} = -1.65$$

$$a = 0.10 \Rightarrow z \text{ critical value} = -1.28$$

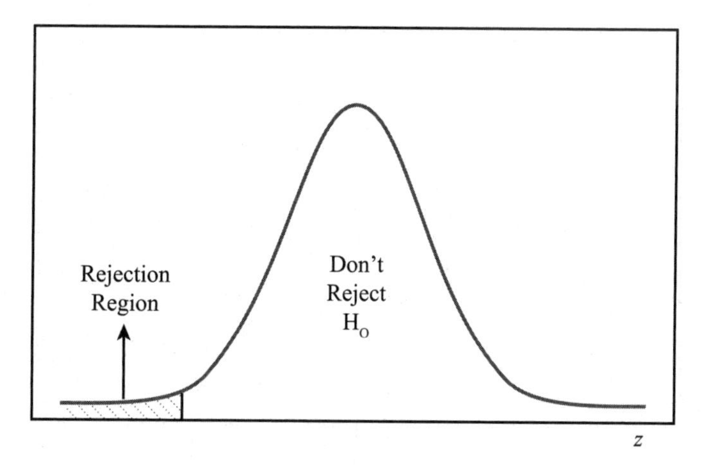

Figure 15.4
One-tail hypothesis test.

Since we have just one rejection region in the left tail of the distribution, if the calculated test statistic falls within this rejection region, then we reject the null hypothesis.

The **one-tail hypothesis test** is used when the alternative hypothesis is being stated as < or >.

Now, let's say our president believes that our seniors have more credit card debt than the national average of $4,100. To test this, we'll use a right tail test. The null and alternative hypotheses are stated as:

$H_0 : \mu \leq \$4,100.$

$H1 : \mu > \$4,100.$

The rejection area in this case is in the right tail of the distribution, which is the shaded area in Figure 15.5. The critical *z*-values in this case are:

$a = 0.01 \Rightarrow z \text{ critical value} = +2.33$

$a = 0.05 \Rightarrow z \text{ critical value} = +1.65$

$a = 0.10 \Rightarrow z \text{ critical value} = +1.28$

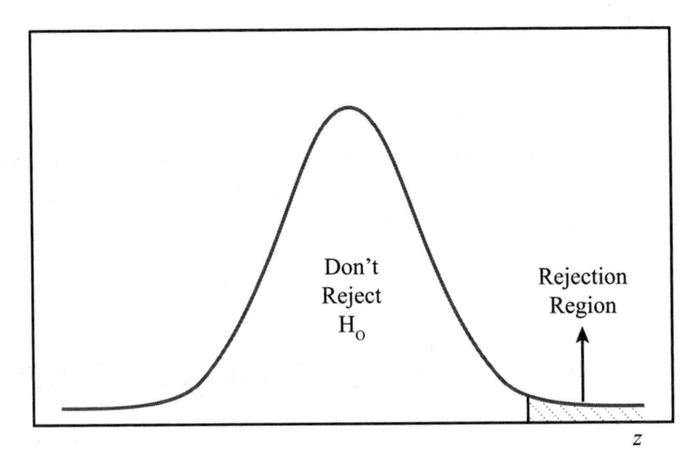

Figure 15.5
One-tail hypothesis test.

As with the previous cases, if the calculated test statistic falls within this rejection region, then we reject the null hypothesis.

BOB'S BASICS

For a one-tail hypothesis test, the rejection region will always be consistent with the direction of the inequality for H_1. For $H_1 : \mu > \$4,100$, the rejection region will be in the right tail of the sampling distribution. For $H_1 : \mu < \$4,100$, the rejection region will be in the left tail.

You need to be careful how you state the null and alternative hypotheses. Your choice will depend on the nature of the test and the motivation of the person conducting it.

If the purpose is to test that the population mean is equal to a specific value, such as our seniors' credit card debt example, assign this statement as the null hypothesis, which results in the following:

$$H_0 : \mu = \$4{,}100$$

$$H_1 : \mu \neq \$4{,}100$$

Often hypothesis testing is performed by researchers who want to prove that their discovery is an improvement over current products or procedures. For example, if Bob invented a golf ball that he claimed would increase your distance off the tee by more than 20 yards, then we would set up the hypotheses as follows:

$$H_0 : \mu \leq 20 \text{ yards}$$

$$H_1 : \mu > 20 \text{ yards}$$

Note that we used the alternative hypothesis to represent the claim that we want to prove statistically so that Bob can make a fortune selling these balls to desperate golfers. Because of this, the alternative hypothesis is also known as the research hypothesis because it represents the position that the researcher wants to establish.

Let's put these concepts to work now and do some hypothesis testing!

Hypothesis Testing for the Population Mean

We'll start with the case when σ, the population standard deviation, is known and then move on to the case when σ is unknown.

When Sigma Is Known

When we use a large sample size ($n \geq 30$) to test our hypothesis, we can rely on our old friend the central limit theorem that we met in Chapter 13. In this case the sampling distribution of the sample means will be normally distributed. If we have a small sample ($n < 30$), then the population from which the samples are drawn must be normally distributed.

To demonstrate this type of hypothesis testing, let's apply it to our college seniors' credit card debt example. We start by setting up our hypotheses as follows:

$$H_0 : \mu = \$4{,}100$$

$$H_1 : \mu \neq \$4{,}100$$

We sample 60 seniors in our college and find that their average credit card debt is $3,900. We'll say that σ, the population standard deviation, is $1,200, and we'll test the hypothesis at $\alpha = 0.05$.

Because the sample size is greater than 30 and we know the value of σ, we calculate the z-test statistic as follows:

$$z = \frac{\bar{x}-\mu}{\sigma_{\bar{x}}} = \frac{3900-4100}{1200/\sqrt{60}} = \frac{-200}{154.92} = -1.29$$

For a two-tail test $\alpha = 0.05$, the critical value = ± 1.96 (shown in Figure 15.6).

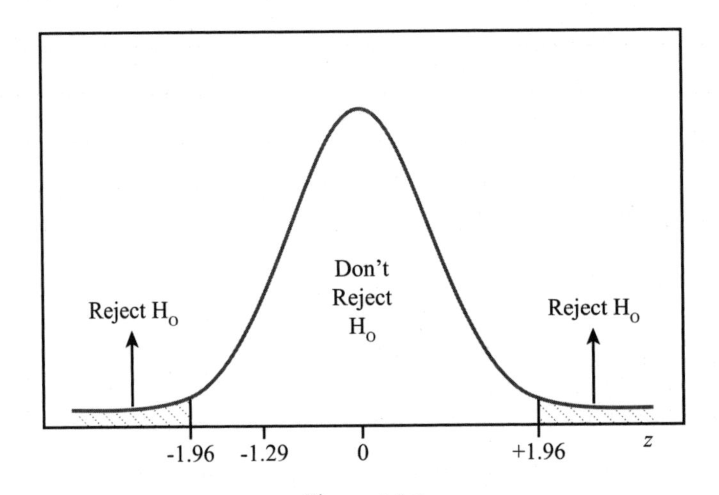

Figure 15.6
A two-tail hypothesis test for the college seniors' credit card debt example.

As you can see in the figure, the calculated z-test statistic of –1.29 falls within the "Don't Reject H_0" region, which allows us to conclude that the average college senior's credit card debt in our college is not significantly different from the national average of $4,100.

When Sigma Is Unknown and Using a Large Sample

Many times, we just don't have enough information to know the value of σ, the population standard deviation. However, as long as our sample size is 30 or more, we can substitute s, the sample standard deviation, for σ. In this case, the only difference will be the calculated z-test statistic in step 3. We will use the following test statistic:

$$z = \frac{\bar{x}-\mu}{s_{\bar{x}}} = \frac{\bar{x}-\mu}{s/\sqrt{n}}$$

To illustrate this technique, let's use the following example.

I don't know about you, but it seems I spend too much time on the phone waiting on hold for a live customer service representative. Let's say a particular company has claimed that the average time a customer waits on hold is less than five minutes. We'll assume we do not know the value of σ. The following table represents the wait time in minutes for a random sample of 30 customers.

Wait Time in Minutes									
6.2	3.8	1.3	5.4	4.7	4.4	4.6	5.0	6.6	8.3
3.2	2.7	4.0	7.3	3.6	4.9	0.5	2.9	2.5	5.6
5.5	4.7	6.5	7.1	4.4	5.2	6.1	7.4	4.8	2.9

Using Excel, we can determine that $\bar{x} = 4.74$ minutes and $s = 1.82$ minutes. At first glance, it appears the company's claim is valid. But let's put it through a hypothesis test with $\alpha = 0.01$ to be sure.

State the hypothesis as:

$$H_0 : \mu \geq 5.0 \text{ minutes}$$

$$H_1 : \mu < 5.0 \text{ minutes}$$

Now, we need to calculate the z-test statistic:

$$z = \frac{\bar{x} - \mu}{s_{\bar{x}}} = \frac{4.74 - 5}{1.82/\sqrt{30}} = -0.78$$

This is a one-tail (left side) hypothesis test with $\alpha = 0.01$, so the critical value is -2.33. Figure 15.7 shows this test graphically.

According to our figure, we do not reject the null hypothesis. In other words, we do not have enough evidence from this sample to support the company's claim that the average wait on hold is less than five minutes. Even though the sample average is actually less than five minutes (4.74), it's too close to five minutes to say there's a difference between the two values. Another way to state this is to say: "The difference between 4.74 and 5.0 is not statistically significant in this case."

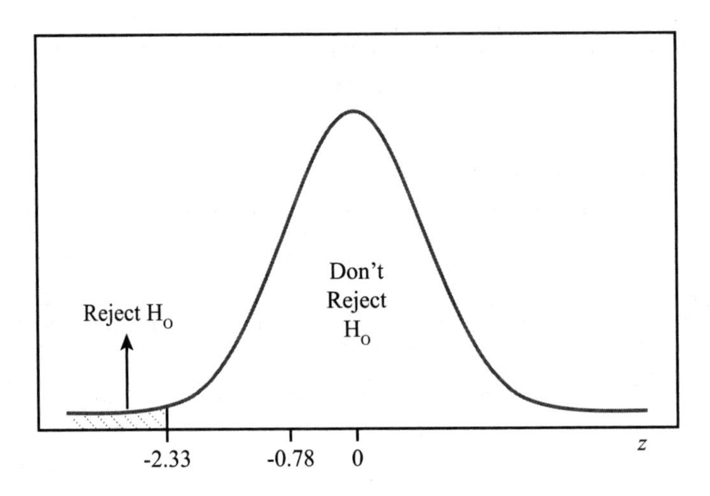

Figure 15.7
A one-tail hypothesis test for the waiting on hold example.

When Sigma Is Unknown and Using a Small Sample

As we did in Chapter 14, when σ is unknown for a small sample size taken from a normally distributed population, we use the Student's t-distribution. In this case, both steps 3 and 4 will be slightly altered. In step 3, the calculated t-test statistic will be:

$$t = \frac{\bar{x}-\mu}{s_{\bar{x}}} = \frac{\bar{x}-\mu}{s/\sqrt{n}}$$

In step 4, we get the critical value(s) using the t-distribution instead of the z-distribution.

Let's illustrate this with an example. According to the National Association of Colleges and Employers (NACE) September 2014 Salary report, the average starting salary for 2014 college graduates is \$48,707. The president of our college believes that our graduates earn more than this, so he asks me to check it for him. I chose a sample of 20 students of our 2014 graduates and found that the average starting salary is \$50,230 and the standard deviation of this sample is \$5,100. Using this information, I can test the claim with the following hypotheses:

$$H_0 : \mu \leq \$48{,}707$$

$$H_1 : \mu > \$48{,}707$$

We can then determine the calculated t-test statistic using the following equation:

$$t = \frac{\bar{x}-\mu}{s_{\bar{x}}} = \frac{50{,}230-48{,}707}{5{,}100/\sqrt{20}} = 1.34$$

We'll test this hypothesis using $\alpha = 0.05$. To find the corresponding critical t-value, we use Table 4 from Appendix B. Here is an excerpt of this table.

Student's *t*-Distribution Table

One tail α	0.200	0.150	0.100	0.0500	0.0250	0.010	0.005
Two tail α	0.400	0.300	0.200	0.100	0.050	0.020	0.010
Conf lev	0.600	0.700	0.800	0.900	0.950	0.980	0.990
d.f.							
1	1.376	1.963	3.078	6.314	12.706	31.821	63.657
2	1.061	1.386	1.886	2.920	4.303	6.965	9.925
3	0.978	1.250	1.638	2.353	3.182	4.541	5.841
4	0.941	1.190	1.533	2.132	2.776	3.747	4.604
5	0.920	1.156	1.476	2.015	2.571	3.365	4.032
6	0.906	1.134	1.440	1.943	2.447	3.143	3.707
7	0.896	1.119	1.415	1.895	2.365	2.998	3.499
8	0.889	1.108	1.397	1.860	2.306	2.896	3.355
9	0.883	1.100	1.383	1.833	2.262	2.821	3.250
10	0.879	1.093	1.372	1.812	2.228	2.764	3.169
11	0.876	1.088	1.363	1.796	2.201	2.718	3.106
12	0.873	1.083	1.356	1.782	2.179	2.681	3.055
13	0.870	1.079	1.350	1.771	2.160	2.650	3.012
14	0.868	1.076	1.345	1.761	2.145	2.624	2.977
15	0.866	1.074	1.341	1.753	2.131	2.602	2.947
16	0.865	1.071	1.337	1.746	2.120	2.583	2.921
17	0.863	1.069	1.333	1.740	2.110	2.567	2.898
18	0.862	1.067	1.330	1.734	2.101	2.552	2.878
19	0.861	1.066	1.328	1.729	2.093	2.539	2.861
20	0.860	1.064	1.325	1.725	2.086	2.528	2.845

If you recall from Chapter 14, we need to determine the degrees of freedom, which is equal to $n - 1 = 20 - 1 = 19$ for this example. Because this is a one-tail (right side) test, we look under the one tail α row and the $\alpha = 0.05$ column resulting in a critical t-value equal to +1.729, which is underlined. Figure 15.8 shows this test graphically.

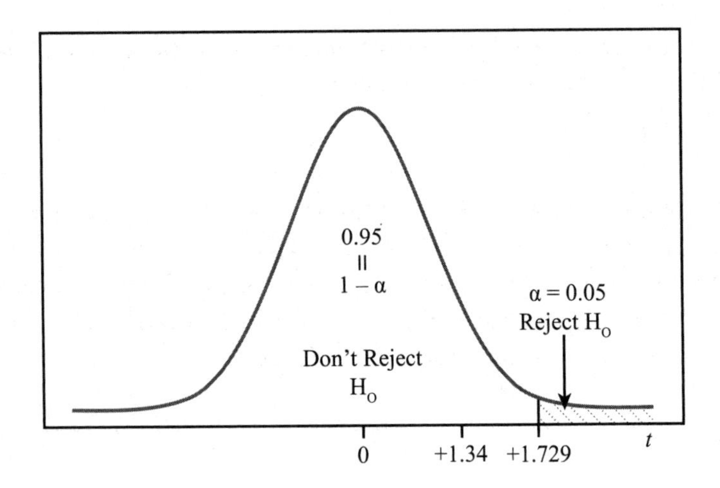

Figure 15.8
Hypothesis test for the average starting salary for college graduates when $\alpha = 0.05$.

BOB'S BASICS

Because this example is a one-tail test on the right side of the distribution, we use a positive critical *t*-value. Had this been a one-tail test on the left side, we would have used a negative critical *t*-value.

As we can see in the previous figure, the calculated *t*-test statistic of 1.34 falls within the white "Don't Reject H_0" region. Therefore, we don't reject H_0 and we conclude that the starting salary for our college graduates is not higher than the national average. Sorry, Mr. President. I could not support your claim!

Let's look at another example to demonstrate a two-tail hypothesis test using the *t*-distribution. I would like to test a claim that the average speed of cars passing a specific spot on the interstate is 65 miles per hour. We can express the hypothesis test as follows:

$$H_0 : \mu = 65 \text{ miles per hour}$$

$$H_1 : \mu \neq 65 \text{ miles per hour}$$

We will assume that we do not know σ and that the speeds follow a normal distribution. The following represents a random sample of the speed of seven cars.

Car Speeds:

62 74 65 68 71 64 68

Using Excel, we can determine that $\bar{x} = 67.43$ mph and $s = 4.16$ *mph* for this sample. We can then determine the calculated t-test statistic as follows:

$$t = \frac{\bar{x} - \mu}{s_{\bar{x}}} = \frac{67.43 - 65}{4.16/\sqrt{7}} = 1.55$$

We'll test this hypothesis using $\alpha = 0.05$. To find the corresponding critical t-value, we use Table 4 from Appendix B. The degrees of freedom for this example equals $n - 1 = 7 - 1 = 6$. Looking at the excerpt of the previous t-table, we can get the critical value. Because this is a two-tail test, we look under the two-tail α row and the $\alpha = 0.05$ column resulting in a critical t-value equal to ±2.447. This test is shown graphically in Figure 15.9.

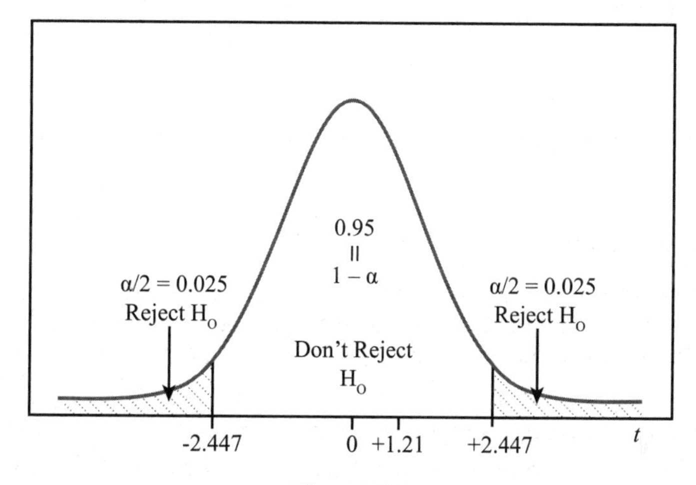

Figure 15.9

Hypothesis test for the car speed example.

As we can see in the previous figure, the calculated t-test statistic of +1.21 falls within the "Do Not Reject H_0" region, so we don't reject H_0. Therefore, there's not enough evidence to conclude that the average speed isn't 65 miles per hour.

The Role of Alpha in Hypothesis Testing

For all the examples in this chapter, we have just stated a value for α, the level of significance. You're probably wondering what impact changing the value of α will have on the hypothesis test. Great question!

In our starting salary for college graduates example, we used $\alpha = 0.05$. Now, if I want to support our college president's claim, it would be in my best interest if I could reject H_0, which would validate his claim. I can do so by choosing a fairly high value for α, say 0.10. In our example, looking at the previous excerpt of the t-distribution and for degrees of freedom = 19, this corresponds to a critical t-value of +1.328 (we are using the right tail of a one-tail hypothesis test). My calculated t-test statistic of +1.34 falls in the "Reject H_0" region, so we reject H_0. This test is shown graphically in Figure 15.10. This means that we can now support our college president's claim that our college graduates' starting salary is higher than the national average.

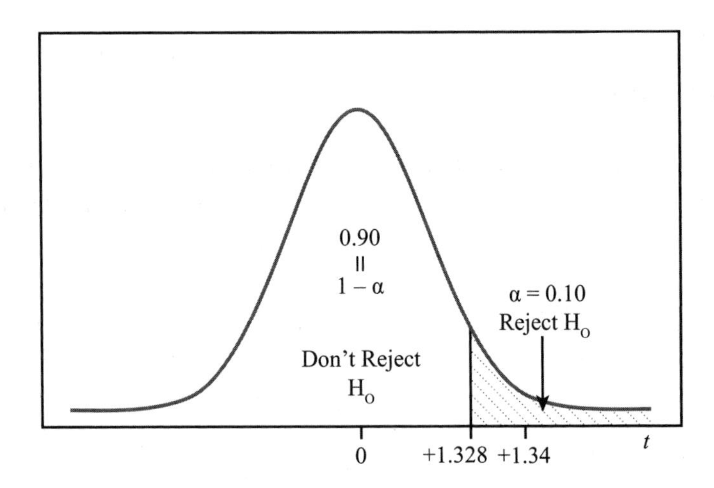

Figure 15.10

The hypothesis test for the average starting salary for college graduates when $\alpha = 0.10$.

However, I must admit I chose a pretty "wimpy" value of $\alpha = 0.10$ in an effort to help prove his claim. In this case, I am willing to accept a 10 percent chance of a Type I error. In general, a hypothesis test that rejects H_0 is most impressive with a low value of α.

The p-Value Method

Just when you thought it was safe to get back in the water, along comes another shark! This is the perfect opportunity to throw another concept at you. You might feel like grumbling a little right now, but in the end you'll be thanking us.

The *p-value* is the smallest level of significance at which the null hypothesis will be rejected, assuming the null hypothesis is true. The p-value is sometimes referred to as the *observed level of significance.* I know this may sound like a lot of mumbo-jumbo right now, but an illustration will help make this clear.

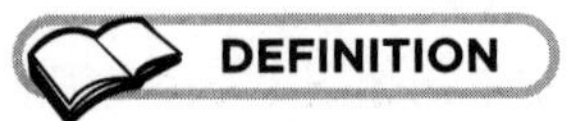

The **observed level of significance** is the smallest level of significance at which the null hypothesis will be rejected, assuming the null hypothesis is true. It is also known as the p-value.

The p-Value for a One-Tail Test

Using the previous example of the average credit card debt for college seniors, let's say I want to test whether our college seniors owe less than the national average. Our hypotheses are as follows:

$$H_0 : \mu \geq \$4{,}100$$

$$H_1 : \mu < \$4{,}100$$

The calculated z-test statistic is -1.29 as before. To get the p-value, we need to get the $P(z<-1.29)$, so we look in the standardized normal z table for $z = -1.29$. This gives me 0.0985. This is the p-value! Yes, it's that easy. This is shown in Figure 15.11.

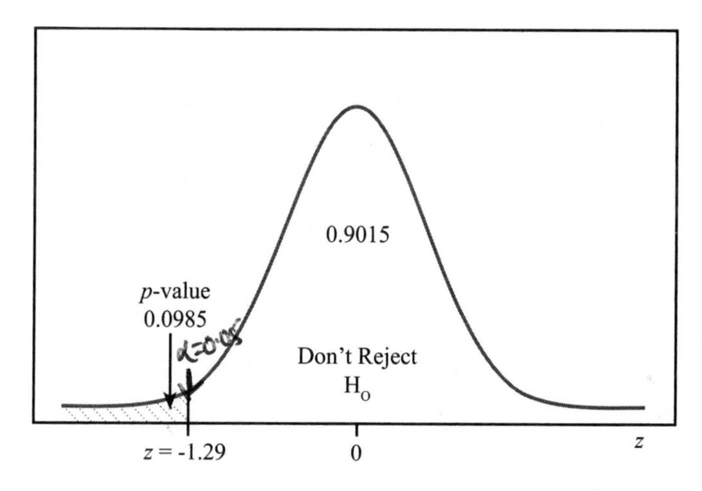

Figure 15.11
The p-value for the college seniors' credit card debt example.

BOB'S BASICS

Note that here we use the value from the table as is since we are looking at the area to the left of the z-value. If we have a right tail test instead, we would have to subtract the area we get from one. For example, if z is +1.29 instead of -1.29, then the p-value $= 1 - 0.9015 = .0985$. If you remember from Chapter 11, $P(z > 1.29) = 1 - P(z < 1.29)$.

Because our p-value of 0.0985 is more than the value of α (set at 0.05), we do not reject H_0. Most statistical software packages (including Excel) provide p-values with the analysis.

Another way to describe this p-value is to say, in a very scholarly voice, "Our results are significant at the 0.0985 level." This means that as long as the value of α is 0.0985 or larger, we will reject H_0, which is normally good news for researchers trying to validate their findings.

BOB'S BASICS

We can use the p-value to determine whether or not to reject the null hypothesis. In general ...

- If p-value $\leq \alpha$, we reject the null hypothesis.
- If p-value $> \alpha$, we do not reject the null hypothesis.

Calculating the p-value for a two-tail hypothesis test is slightly different, and I'll show you how in the next section.

The p-Value for a Two-Tail Test

Recall that you use a two-tail hypothesis test when the null hypothesis is stated as an equality. For example, let's test a claim that states the average number of miles driven by a passenger vehicle in a year equals 11,500 miles. Bob had serious reservations about this claim after spending half the day being a taxi driver to his kids. We would state the hypotheses as follows:

$$H_0 : \mu = 11{,}500 \text{ miles}$$

$$H_1 : \mu \neq 11{,}500 \text{ miles}$$

Let's assume $\sigma = 3{,}000$ miles, and we want to set $\alpha = 0.05$. We sample 80 drivers and determine the average number of miles driven is 11,900. What is our p-value, and what do we conclude about the hypothesis?

We can get the calculated z-test statistic as follows:

$$z = \frac{\bar{x} - \mu}{\sigma_{\bar{x}}} = \frac{11{,}900 - 11{,}500}{3{,}000/\sqrt{80}} = +1.19$$

The shaded area in Figure 15.12 shows the p-value for this test.

According to Table 3 in Appendix B, the $P(z \leq +1.19) = 0.8830$. This means the shaded region in the right tail of Figure 15.12 is $P(z > +1.19) = 1 - 0.8830 = 0.117$. Because this is a two-tail test, we need to double this area to arrive at our p-value. According to our figure, the p-value is the total area of both shaded regions, which is $2 \times 0.117 = 0.234$. Because the p-value $> \alpha$, we do not reject the null hypothesis. Our data supports the claim that the average number of miles driven per year by a passenger vehicle is 11,500.

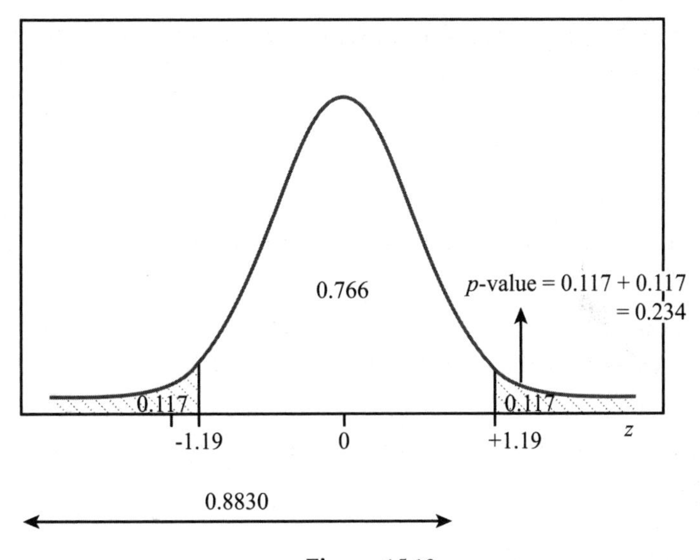

Figure 15.12

The p-value for the miles driven per year example.

In general, the smaller the p-value, the more confident we are about rejecting the null hypothesis. In most cases a researcher is attempting to find support for the alternative hypothesis. A low p-value provides support that brings joy to his or her heart.

BOB'S BASICS

It is not possible to determine the p-value for a hypothesis test when using the Student's t-distribution table in Appendix B. However, most statistical software will provide the p-value as part of the standard analysis. We can also use Excel to get the p-value as shown in the next section.

Using Excel for Hypothesis Testing

We can generate the p-value using Excel functions. For a z-distribution, we use Excel's NORM.S.DIST function and for a t-distribution, we use Excel's T.DIST or T.DIST.RT or T.DIST.2T functions. Let's see examples for each one of them.

Using Excel's NORM.S.DIST Function

Use Excel's NORM.S.DIST function to find the p-value when you are using the z-distribution. The function has the following characteristics:

NORM.S.DIST(z, cumulative)

where:

z = the calculated z-test statistic

cumulative = TRUE, to get the area to the left of the z-value

For instance, we would use this function in the previous example of the average credit card debt for college seniors with the following hypotheses:

$$H_0 : \mu \geq \$4{,}100$$

$$H_1 : \mu < \$4{,}100$$

The calculated z-test statistic is -1.29 as before. Figure 15.13 shows the NORM.S.DIST function being used to determine the p-value for this example, which is a one-tail test. As you can see from the figure, the Excel function gives us a p-value of 0.0985. This is the same value we obtained when we used the table.

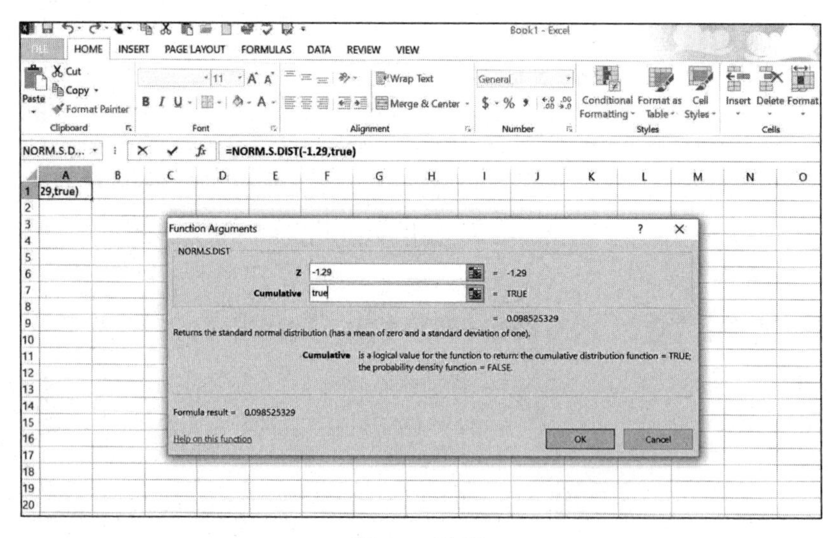

Figure 15.13
Excel's NORM.S.DIST function for a one-tail test.

Using Excel's T.DIST, T.DIST.RT, and T.DIST.2T Functions

Use these functions to find the p-value when you are using the t-distribution. Excel's T.DIST function gives you the area to the left of the t-value, while Excel's T.DIST.RT function gives you the area to the right of the t-value. For two-tail testing, use Excel's T.DIST.2T function.

Excel's T.DIST functions have the following characteristics:

T.DIST(x,degrees_of_freedom)

where:

x = the calculated t-test statistic

For instance, we would use the T.DIST.RT function in the previous example of our graduates' average starting salary with the following hypotheses:

$$H_0 : \mu \leq \$48,707$$

$$H_1 : \mu > \$48,707$$

Figure 15.14 shows the T.DIST.RT function being used to determine the p-value for this example where $t = 1.34$, which is a right tail test.

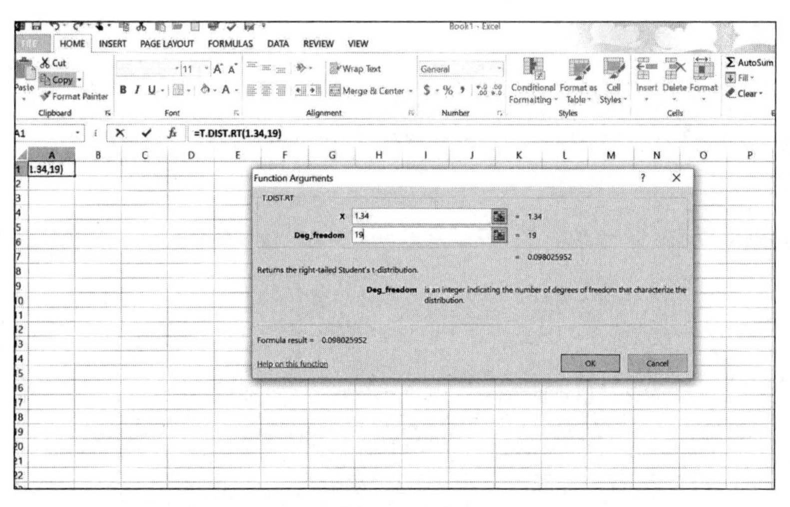

Figure 15.14
Excel's T.DIST.RT function for a one-tail test.

For a two-tail test, we would use the T.DIST.2T function, like in the previous example of the average speed of cars with the following hypotheses:

$$H_0 : \mu = 65 \text{ miles per hour}$$

$$H_1 : \mu \neq 65 \text{ miles per hour}$$

Figure 15.15 shows the T.DIST.2T function being used to determine the p-value for this example where $t = 1.21$, which is a two-tail test.

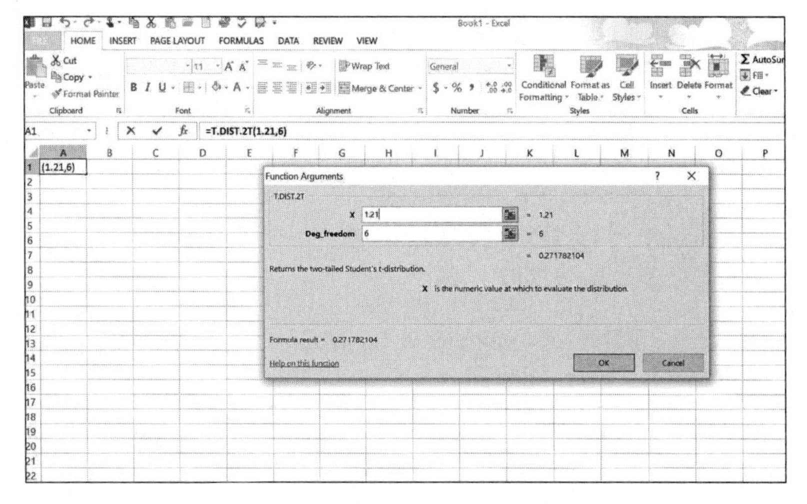

Figure 15.15

Excel's T.DIST.2T function for a two-tail test.

Using Excel's T.INV and T.INV.2T Functions

We can also use Excel to get the critical t-value using Excel's T.INV and T.INV.2T functions. Excel's T.INV functions have the following characteristics:

T.INV(probability,deg_freedom)

Where:

probability = the probability associated with the Student's t-distribution

For instance, Figure 15.16 shows the T.INV.2T function being used to determine the critical t-value for $\alpha = 0.05$ and $d.f. = 6$ from our average speed of cars example, which is a two-tail test. It has the following hypotheses:

$$H_0 : \mu = 65 \text{ miles per hour}$$

$$H_1 : \mu \neq 65 \text{ miles per hour}$$

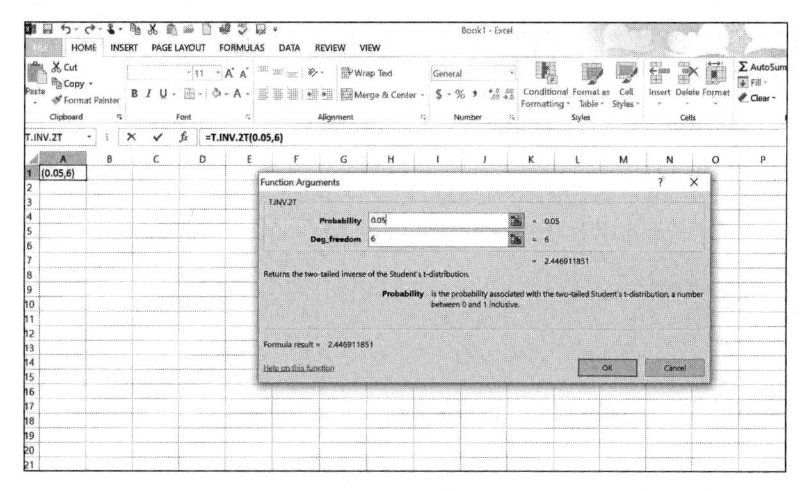

Figure 15.16

Excel's T.INV.2T function for a two-tail test.

As you can see from the figure, the Excel function =T.INV.2T(0.05, 6) gives us a critical value of 2.447. This is the same value we obtained from the table in our previous average speed of cars example.

Excel's T.INV function gives the critical t-value for a left-side test. Figure 15.17 shows the T.INV function being used to determine the critical t-value for $\alpha = 0.01$ and $d.f. = 24$ for a left tail test.

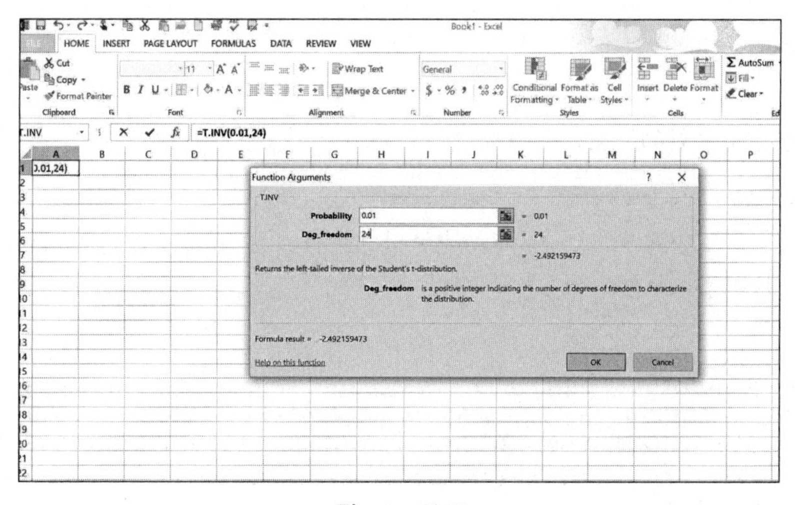

Figure 15.17

Excel's T.INV function for a left tail test.

The critical t-value is the same for a right-tail test, except positive instead of negative.

Hypothesis Testing for the Population Proportion with Large Samples

You can perform hypothesis testing for the proportion of a population as long as the sample size is large enough. Recall from Chapter 13 that proportion data follows the binomial distribution, which can be approximated by the normal distribution under the following conditions:

$$np \geq 5 \text{ and } nq \geq 5$$

where:

p = the probability of a success in the population

q = the probability of a failure in the population ($q = 1 - p$)

We will examine both one-tail and two-tail hypothesis testing for the proportion in the following sections.

One-Tail Hypothesis Test for the Proportion

Let's say we would like to test the hypothesis that more than 30 percent of U.S. households have wireless internet access. We would state the hypotheses as:

$$H_0 : p \leq 0.30$$

$$H_1 : p > 0.30$$

Where:

p = the proportion of U.S. households with wireless internet access.

We collect a sample of 150 households and find that 57 of them have wireless internet access. What can we conclude at the $\alpha = 0.05$ level?

Be careful not to confuse this definition of p—the proportion—with the p-value that we talked about earlier.

Our first step is to calculate σ_p, the standard error of the proportion, which was described in Chapter 13 using the following equation:

$$\sigma_p = \sqrt{\frac{p(1-P)}{n}}$$

where p = the population proportion, which is the one assumed by the null hypothesis. For our example:

$$\sigma_p = \sqrt{\frac{(0.30)(1-0.30)}{150}} = 0.037$$

We need to determine the sample proportion $\overline{p}$, as follows:

$$\overline{p} = \frac{x}{n} = \frac{57}{150} = 0.38$$

Next, we can determine the calculated z-test statistic using:

$$z = \frac{\overline{p} - p}{\sigma_p}$$

For our example:

$$z = \frac{\overline{p} - p}{\sigma_p} = \frac{0.38 - 0.3}{0.037} = 2.16$$

The critical z-value for a one-tail test with $\alpha = 0.05$ is +1.65. This hypothesis test is shown graphically in Figure 15.18.

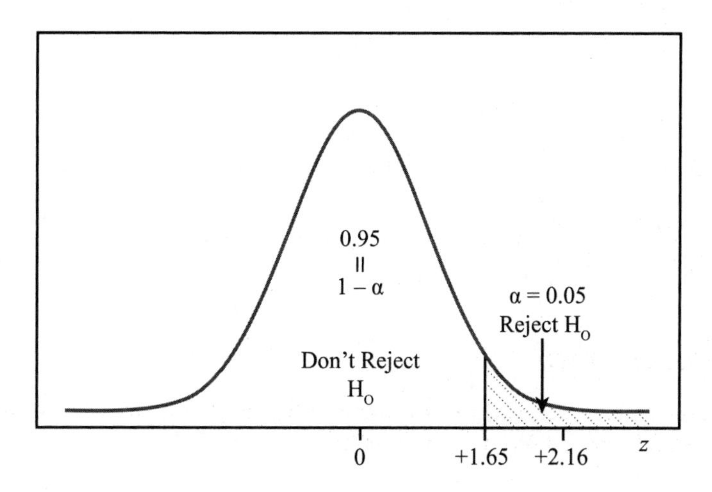

Figure 15.18

Hypothesis test for the wireless internet access example.

As you can see in Figure 15.18, the calculated z-test statistic of +2.16 is within the "Reject H_0" region. Therefore, we conclude that the proportion of U.S. households with wireless internet access exceeds 30 percent.

We can also get the p-value for this test using our standardized normal z table (Table 3 in Appendix B) as follows:

$$P(z > +2.16) = 1 - P(z \leq +2.16) = 1 - 0.9846 = 0.0154$$

Therefore, our results are significant at the 0.0154 level. As long as $\alpha = 0.0154$, we will be able to reject H_0.

Two-Tail Hypothesis Test for the Proportion

We'll wrap this chapter up with one final two-tail example. Here, we want to test a hypothesis for a company that claims 50 percent of their customers are male. We state our hypotheses as:

$$H_0 : p = 0.50$$

$$H_1 : p \neq 0.50$$

We randomly select 256 customers and find that 47 percent are male. What can we conclude at the $\alpha = 0.05$ level?

We need to determine σ_p, the standard error of the proportion:

$$\sigma_p = \sqrt{\frac{p(1-P)}{n}} = \sqrt{\frac{(0.5)(1-0.5)}{256}} = 0.0312$$

Next, we can determine the calculated z-test statistic:

$$z = \frac{\bar{p}-p}{\sigma_p} = \frac{0.47-0.5}{0.0312} = -0.96$$

The critical z-value for a two-tail test with $\alpha = 0.05$ is ± 1.96. This hypothesis test is shown graphically in Figure 15.19.

As you can see in Figure 15.19, the calculated z-test statistic of -0.96 is within the "Do Not Reject H_0" region. Therefore, we conclude that the proportion of male customers is not significantly different from 50 percent for this company.

We can also calculate the p-value for this test using our standardized normal z table (Table 3 in Appendix B) as follows:

$$P(z \leq -0.96) = 0.1685$$

Because this is a two-tail test, the p-value would be $2 \times 0.1685 = 0.337$. Since the p-value is greater than α, we don't reject H_0.

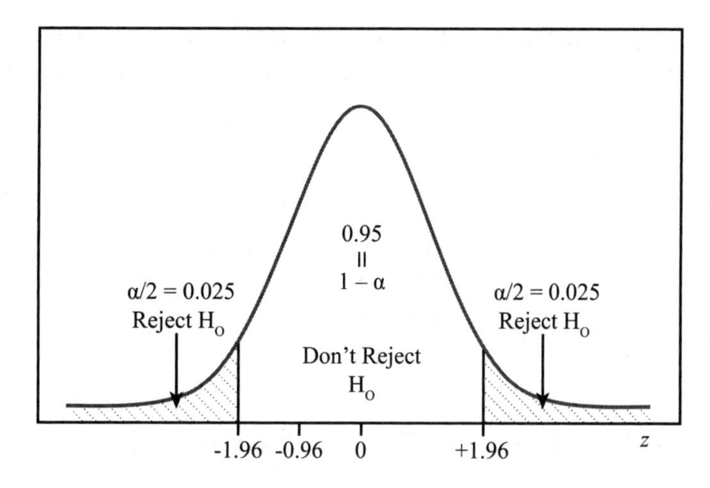

Figure 15.19
Hypothesis test for the percentage of males example.

Practice Problems

1. Formulate a hypothesis statement for the following claim: "The average age of our customers is less than 40 years old." A sample of 50 customers had an average age of 38.7 years. Assume the population standard deviation is 12.5 years. Using $\alpha = 0.05$, test your hypothesis. What is your conclusion?

2. Formulate a hypothesis statement for the following claim: "The average life of our light bulbs is more than 1,000 hours." A sample of 32 light bulbs had an average life of 1,190 hours. Assume the population standard deviation is 325 hours. Using $\alpha = 0.02$, test your hypothesis. What is your conclusion?

3. Formulate a hypothesis statement for the following claim: "The average delivery time is less than 30 minutes." A sample of 42 deliveries had an average time of 26.9 minutes. Assume the population standard deviation is 8 minutes. Using $\alpha = 0.01$, test your hypothesis. What is your conclusion?

4. Formulate a hypothesis statement for the following claim: "Students graduating from college have an average student loan debt of $32,700." A sample of 40 college graduates averaged $32,450 in student loan debt. Assume the population standard deviation is $950. Using $\alpha = 0.05$, test your hypothesis. What is your conclusion?

5. Test the claim that the average SAT score for graduating high school students is equal to 1500. A random sample of 70 students was selected, and the average SAT score was 1435. Assume $\sigma = 310$ and use $\alpha = 0.10$. What is the p-value for this sample?

6. A student organization at a small business college claims that the average class size is greater than 35 students. Test this claim at $\alpha = 0.02$, using the following sample of class size:

42	28	36	47	35	41	33	30	39	48

   Assume the population is normally distributed and that σ is unknown.

7. Test the claim that the average gasoline consumption per car in the United States is more than 7 liters per day. (We're going metric here!) Use the random sample below, which represents daily gasoline usage for one car:

9	6	4	12	4	3	18	10	4	5
3	8	4	11	3	5	8	4	12	10
9	5	15	17	6	13	7	8	14	9

   Assume the population is normally distributed, and that σ is unknown. Use $\alpha = 0.05$ and determine the p-value for this sample.

8. Test the claim that the proportion of Republican voters in a particular city is less than 40 percent. A random sample of 175 voters was selected and found to consist of 30 percent Republicans. Use $\alpha = 0.01$ and determine the p-value for this sample.

9. Test the claim that the proportion of teenage cell phone users exceeding their allotted monthly minutes equals 65 percent. A random sample of 225 teenagers was selected and found to consist of 69 percent exceeding their minutes. Use $\alpha = 0.05$ and determine the p-value for this sample.

10. Test the claim that the mean number of hours that undergraduate students work at a particular college is less than 15 hours per week. A random sample of 60 students was selected, and the average number of working hours was 13.5 hours per week. Assume $\sigma = 5$ hours, and use $\alpha = 0.10$. What is the p-value for this sample?

The Least You Need to Know

- The null hypothesis, denoted by H_0, represents the status quo and involves stating the belief that the mean of the population is $\leq$, $=$, or $\geq$ a specific value.
- The alternative hypothesis, denoted by H_1, represents the opposite of the null hypothesis and holds true if the null hypothesis is found to be false.
- Use a two-tail hypothesis test whenever the alternative hypothesis is expressed as $\neq$; whereas a one-tail hypothesis test involves the alternative hypothesis being stated as $<$ or $>$.

- A Type I error occurs when the null hypothesis is rejected when, in reality, it is true. The probability of this error occurring is known as α, the level of significance.
- A Type II error occurs when the null hypothesis is not rejected when, in reality, it is not true. The probability of this error occurring is known as β.
- The smaller the value of α, the level of significance, the more difficult it is to reject the null hypothesis.
- The p-value is the smallest level of significance at which the null hypothesis will be rejected, assuming the null hypothesis is true.
- If the p-value $\leq \alpha$, we reject the null hypothesis. If p-value $> \alpha$, we do not reject the null hypothesis.
- Use the Student's t-distribution for the hypothesis test when $n < 30$, σ is unknown, and the population is normally distributed.

CHAPTER

16

Hypothesis Testing with Two Populations

In This Chapter

- Developing the sampling distribution for the difference in means
- Testing the difference in means between populations when the population standard deviations (σ_1 and σ_2) are known and when they are unknown
- Distinguishing between independent and dependent samples
- Using Excel to perform a hypothesis test
- Testing the difference in proportions between populations

Now we're really cooking. Because you have done so well with one sample hypothesis testing, you are ready to graduate to the next level—two-sample testing. Here we often test to see whether there is a difference between two separate populations. For instance, we could test to see whether there was a difference between the average golf score for Bob's sons: Brian and John. But as an "experienced" parent, Bob knew better than to go near that one.

Because many similarities exist between the concepts of this chapter and those of Chapter 15, you should have a firm handle on the previous chapter's material before you jump into this one.

The Concept of Testing Two Populations

Many statistical studies involve comparing the same parameter, such as a mean, between two different populations. For example:

- Is there a difference in average SAT scores between males and females?
- Do "long-life" light bulbs really outlast standard light bulbs?
- Does the average selling price of a house in Newark differ from the average selling price for a house in Wilmington?
- Is there a difference between registered nurses' salaries in New York and in California?

To answer such questions, we need to explore a new sampling distribution. (I promise this will be the last.) This one has the fanciest name of them all—the *sampling distribution for the difference in means.* (Dramatic background music brings us to the edge of our seats.)

The **sampling distribution for the difference in means** describes the probability of observing various intervals for the difference between two sample means.

Sampling Distribution for the Difference in Means

Just like the sampling distribution for the sample mean we had in Chapter 13, we can get the sampling distribution for the difference in means, which can best be described in Figure 16.1.

As an example, let's consider testing for a difference in the SAT scores for students between two states: North Carolina and South Carolina. We'll assign students in North Carolina as Population 1 and South Carolina as Population 2. Graph 1 in Figure 16.1 represents the distribution of the SAT scores for North Carolina students with mean μ_1 and standard deviation σ_1. Similarly, Graph 2 represents the same for South Carolina with μ_2 and σ_2, respectively.

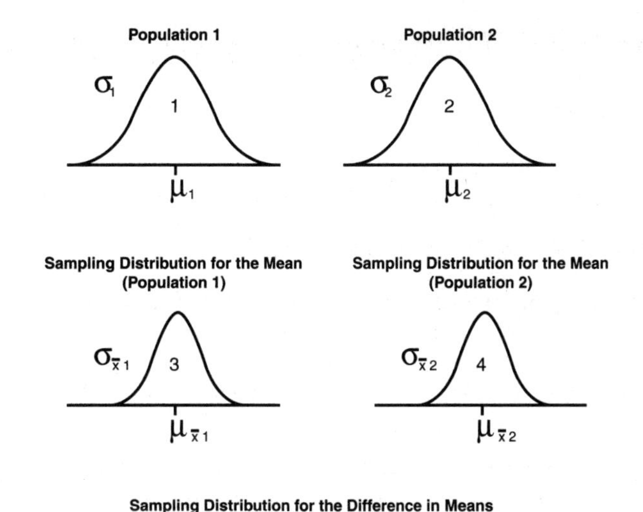

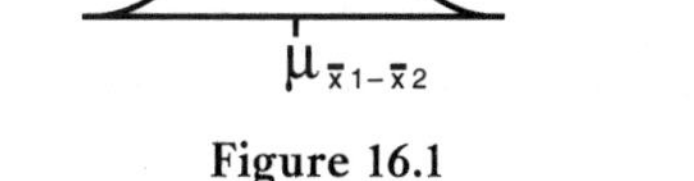

Figure 16.1
The sampling distribution for the difference in means.

Graph 3 represents the sampling distribution for the mean for the North Carolina students. This graph is the result of taking samples of size n_1 and plotting the distribution of sample means. Recall that we discussed this distribution of sample means back in Chapter 13. The mean of this distribution would be:

$$\mu_{\bar{x}_1} = \mu_1$$

This is according to the Central Limit Theorem from Chapter 13. The same logic holds true for Graph 4 for the South Carolina population with mean of $\mu_{\bar{x}2} = \mu_2$.

Graph 5 in Figure 16.1 shows the distribution that represents the difference of sample means from North Carolina and South Carolina populations. This is the sampling distribution for the difference in means, which has the following mean:

$$\mu_{\bar{x}_1 - \bar{x}_2} = \mu_{\bar{x}_1} - \mu_{\bar{x}_2}$$

In other words, the mean of this distribution, shown in Graph 5, is the difference between the means of Graphs 3 and 4.

The standard deviation for Graph 5 is known as the *standard error of the difference between two means* and is calculated with:

$$\sigma_{\bar{x}_1-\bar{x}_2}=\sqrt{\frac{\sigma_1^2}{n_1}+\frac{\sigma_2^2}{n_2}}$$

where:

σ_1^2, σ_2^2 = the variance for Populations 1 and 2

n_1, n_2 = the sample size from Populations 1 and 2

DEFINITION

The **standard error of the difference between two means** describes the variation in the difference between two sample means and is calculated using $\sigma_{\bar{x}_1-\bar{x}_2}=\sqrt{\frac{\sigma_1^2}{n_1}+\frac{\sigma_2^2}{n_2}}$.

Now before you pull the rest of your hair out, let's put these guys to work in the following section.

Testing for the Differences Between Two Population Means with Independent Samples

For this hypothesis test, we assume that the two samples are independent of each other. Samples are independent if they are not related in any way to each other. We are going to start with the case when the population standard deviations, σ_1 and σ_2, are known and then move on to the case when they are unknown.

When the Population Standard Deviations Are Known

When the sample sizes from both populations of interest are greater than 30, the central limit theorem allows us to use the normal distribution to approximate the sampling distribution for the difference in means. Let's demonstrate this technique with the example we started in the previous section. We want to test whether there's a difference in the average SAT scores for students in North Carolina and in South Carolina. To investigate that, we selected two random samples from the two states. A random sample of 75 students from North Carolina shows an average SAT score of 1,480. A random sample of 70 students from South Carolina shows an average SAT score of 1,450. Assume that the population standard deviation for the SAT scores in North Carolina is 200 and in South Carolina is 150. Using $\alpha = 0.05$, we want to test whether there is a difference in the SAT scores between these two states. How do we do that? Read on!

We are going to follow the same 5 steps we used in Chapter 15.

Step 1: State the null and alternative hypotheses.

This is a two-tail test since we are testing whether there is a difference between μ_1 and μ_2. The null hypothesis states that there is no difference, $\mu_1 = \mu_2$. (Another way to write this is $\mu_1 - \mu_2 = 0$.) The same holds for H_1. So we can write our hypotheses like this:

$$H_0 : \mu_1 - \mu_2 = 0$$

$$H_1 : \mu_1 - \mu_2 \neq 0$$

Step 2: Determine the level of significance: we are using $\alpha = 0.05$.

Step 3: Calculate the z-test statistic:

The calculated z-test statistic is determined by the following equation:

$$z = \frac{(\bar{x}_1 - \bar{x}_2) - (\mu_1 - \mu_2)}{\sigma_{\bar{x}_1 - \bar{x}_2}} = \frac{(\bar{x}_1 - \bar{x}_2) - (\mu_1 - \mu_2)}{\sqrt{\frac{\sigma_1^2}{n_1} + \frac{\sigma_2^2}{n_2}}}$$

We start with the standard error of the difference between two means as follows:

$$\sigma_{\bar{x}_1 - \bar{x}_2} = \sqrt{\frac{\sigma_1^2}{n_1} + \frac{\sigma_2^2}{n_2}} = \sqrt{\frac{(200)^2}{75} + \frac{(150)^2}{70}} = 29.2363$$

We are now ready to determine the calculated z-test statistic as follows:

$$z = \frac{(\bar{x}_1 - \bar{x}_2) - (\mu_1 - \mu_2)}{\sqrt{\frac{\sigma_1^2}{n_1} + \frac{\sigma_2^2}{n_2}}} = \frac{(1{,}480 - 1{,}450) - 0}{29.2363} = 1.03$$

BOB'S BASICS

The term $(\mu_1 - \mu_2)$ refers to the hypothesized difference between the two population means. When the null hypothesis is testing that there is no difference between population means, then the term $(\mu_1 - \mu_2)$ is set to 0.

Step 4: Determine the critical values

The critical z-values for a two-tail test with $\alpha = 0.05$ are ± 1.96. Figure 16.2 shows the results of this hypothesis test.

Step 5: State your decision.

According to Figure 16.2, the calculated z-test statistic of 1.03 falls within the "Don't Reject H_0" region, which leads us to conclude that there's no significant difference between the average SAT scores for North Carolina and South Carolina.

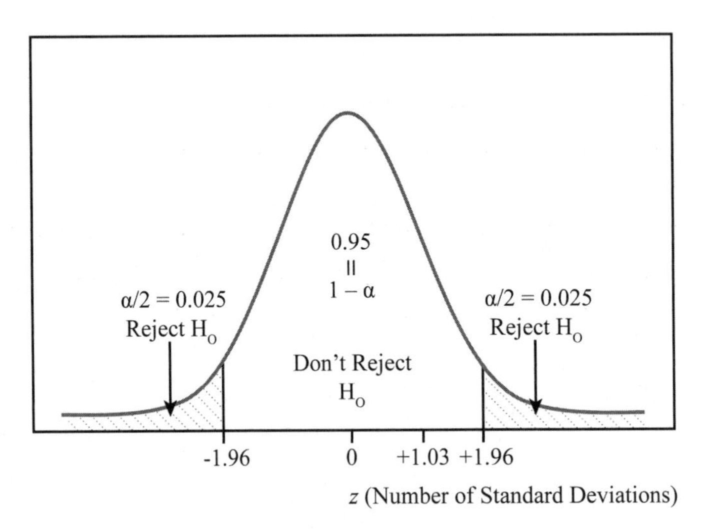

Figure 16.2
Hypothesis test for the SAT example.

As we did in Chapter 15, we can also find the *p*-value for this example by using the standard normal *z* table found in Appendix B as follows:

$$P(z > +1.03) = 1 - P(z \leq +1.03) = 1 - 0.8485 = 0.1515$$

Because this is a two-tail test, we need to double this area to arrive at our *p*-value, which is $2 \times 0.1515 = 0.303$. Because the *p*-value > α, we do not reject the null hypothesis.

This technique applies to large samples, but what if we have small samples? We can also apply this technique to hypothesis testing that involves sample sizes less than 30 as long as both populations are normally distributed.

Testing a Difference Other Than Zero

In the previous example, we were just testing whether or not there was any difference between the two populations. We can also test whether the difference exceeds a certain value. As an example, suppose we want to test the hypothesis that the average salary of a mathematician in Florida exceeds the average salary of a mathematician in Texas by more than $5,000. We would state the hypotheses as follows:

$$H_0 : \mu_1 - \mu_2 \leq 5{,}000$$

$$H_1 : \mu_1 - \mu_2 > 5{,}000$$

where:

μ_1 = the mean salary of a mathematician in Florida

μ_2 = the mean salary of a mathematician in Texas

We'll assume that σ_1 = \$8,100 and σ_2 = \$7,600, and we'll test this hypothesis at the $\alpha = 0.05$ level.

A sample of 42 mathematicians from Florida had a mean salary of \$85,500, whereas a sample of 54 mathematicians from Texas had a mean salary of \$76,000.

The standard error of the difference between two means is:

$$\sigma_{\bar{x}_1-\bar{x}_2} = \sqrt{\frac{\sigma_1^2}{n_1}+\frac{\sigma_2^2}{n_2}} = \sqrt{\frac{(8{,}100)^2}{42}+\frac{(7{,}600)^2}{54}} = \$1{,}622.30$$

Our calculated z-test statistic becomes:

$$z = \frac{(\bar{x}_1-\bar{x}_2)-(\mu_1-\mu_2)}{\sqrt{\frac{\sigma_1^2}{n_1}+\frac{\sigma_2^2}{n_2}}} = \frac{(85{,}500-76{,}000)-(5{,}000)}{1{,}622.3} = 2.77$$

The results of this hypothesis test are shown in Figure 16.3.

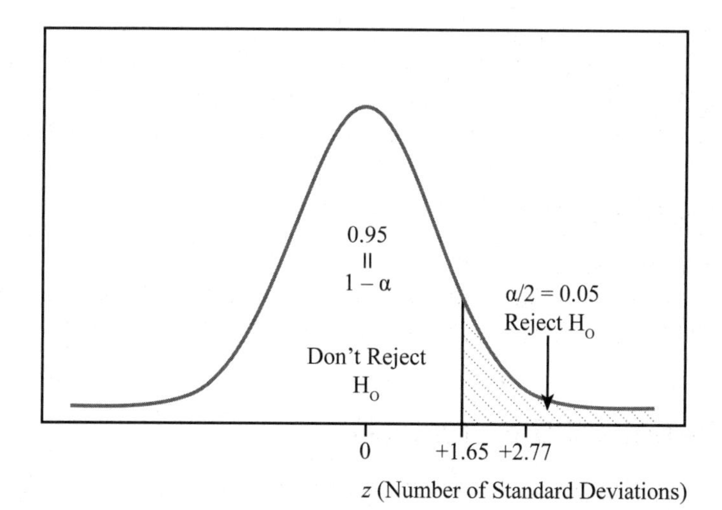

Figure 16.3
The hypothesis test for the mathematicians' salary example.

The critical z-value for a one-tail (right side) test with $\alpha = 0.05$ is +1.65. According to Figure 16.3, this places the calculated z-test statistic of +2.77 in the "Reject H_0" region, which leads to our conclusion that the difference in salaries between the two states exceeds \$5,000.

When the Population Standard Deviations Are Unknown

In many cases, the population standard deviation, σ, is not known. So what can we do? I know that by now you are familiar—we made the same adjustments for small sample sizes back in Chapters 14 and 15. We have to make two changes to the previous technique:

1. We use the sample standard deviation, s, to approximate the population standard deviation, σ.
2. We use the t-distribution instead of the z-distribution. However, if one or both of our sample sizes is less than 30, then the population needs to be normally distributed to use any of the following techniques.

In this case when the population standard deviations are unknown, we have two cases to consider:

- If the population variances are not equal.
- If the population variances are equal.

As you'll see below, the equation for the standard error of the difference between the two means, $\sigma_{\bar{s}_1-\bar{s}_2}$, will differ with each case. Let's look at each instance in detail.

Unequal Population Variances When the Population Standard Deviations Are Unknown

In this case we are assuming that the population standard deviations are not equal, $\sigma_1 \neq \sigma_2$. As we did in Chapter 15, we can follow the same steps but with the two changes in steps 3 and 4.

In step 3, the standard error of the difference between two means is as follows:

$$\sigma_{\bar{s}_1-\bar{s}_2} = \sqrt{\frac{s_1^2}{n_1}+\frac{s_2^2}{n_2}}$$

The calculated t-test statistic will be:

$$t = \frac{(\bar{x}_1-\bar{x}_2)-(\mu_1-\mu_2)}{\sqrt{\frac{s_1^2}{n_1}+\frac{s_2^2}{n_2}}}$$

In step 4, we get the critical value(s) using the t-distribution instead of the z-distribution. The sampling distribution for the difference between sample means for this scenario follows the Student's t-distribution.

Let's illustrate this with an example. According to the Bureau of Labor Statistics, in May 2011 California and Massachusetts were among the top paying states for registered nurses. Let's say I want to check whether there's a difference in the annual salary for registered nurses between California and Massachusetts. I selected a random sample of 25 registered nurses from California and found their average salary is $90,800 with a standard deviation of $9,050. I selected another random sample of 20 registered nurses from Massachusetts and found their average salary is $86,400 with a standard deviation of $8,020. Assuming that the annual salary for registered nurses in both states are normally distributed, to investigate this issue, I start with my hypotheses:

$$H_0 : \mu_1 - \mu_2 = 0$$

$$H_1 : \mu_1 - \mu_2 \neq 0$$

Where:

μ_1 = the mean annual salary for registered nurses in California.

μ_2 = the mean annual salary for registered nurses in Massachusetts.

Next, I'm going to calculate the standard error of the difference between two means as follows:

$$\sigma_{\bar{x}_1 - \bar{x}_2} = \sqrt{\frac{s_1^2}{n_1} + \frac{s_2^2}{n_2}} = \sqrt{\frac{(9{,}050)^2}{25} + \frac{(8{,}020)^2}{20}} = 2{,}547.96$$

We are now ready to determine the calculated t-test statistic as follows:

$$t = \frac{(\bar{x}_1 - \bar{x}_2) - (\mu_1 - \mu_2)}{\sqrt{\frac{s_1^2}{n_1} + \frac{s_2^2}{n_2}}} = \frac{(90{,}800 - 86{,}400) - 0}{2{,}547.96} = 1.73$$

Now, here is something different from the previous cases. To determine the degrees of freedom for the t-distribution, we use the following equation in this case (hold on to your hat):

$$d.f. = \frac{\left(\frac{s_1^2}{n_1} + \frac{s_2^2}{n_2}\right)^2}{\frac{\left(\frac{s_1^2}{n_1}\right)^2}{n_1 - 1} + \frac{\left(\frac{s_2^2}{n_2}\right)^2}{n_2 - 1}}$$

Before you have a seizure, let me demonstrate that this animal's bark is worse than its bite. In our registered nurses' salary example, we start with the following terms

$$\frac{s_1^2}{n_1} = \frac{(9{,}050)^2}{25} = 3{,}276{,}100 \text{ and } \frac{s_2^2}{n_2} = \frac{(8{,}020)^2}{20} = 3{,}216{,}020$$

We just substitute these values into the previous *d.f.* equation:

$$d.f.=\frac{\left(\frac{s_1^2}{n_1}+\frac{s_2^2}{n_2}\right)^2}{\frac{\left(\frac{s_1^2}{n_1}\right)^2}{n_1-1}+\frac{\left(\frac{s_2^2}{n_2}\right)^2}{n_2-1}}=\frac{(3{,}276{,}100+3{,}216{,}020)^2}{\frac{(3{,}276{,}100)^2}{25-1}+\frac{(3{,}216{,}020)^2}{20-1}}=42.5064$$

42.56 falls between 40 and 45. Some books will round to the nearest number. To be on the conservative side, though, we are going to round down to 40. At $\alpha = 0.01$ and degrees of freedom = 40, the critical *t*-value = ±2.704. The results of this hypothesis test are shown in Figure 16.4.

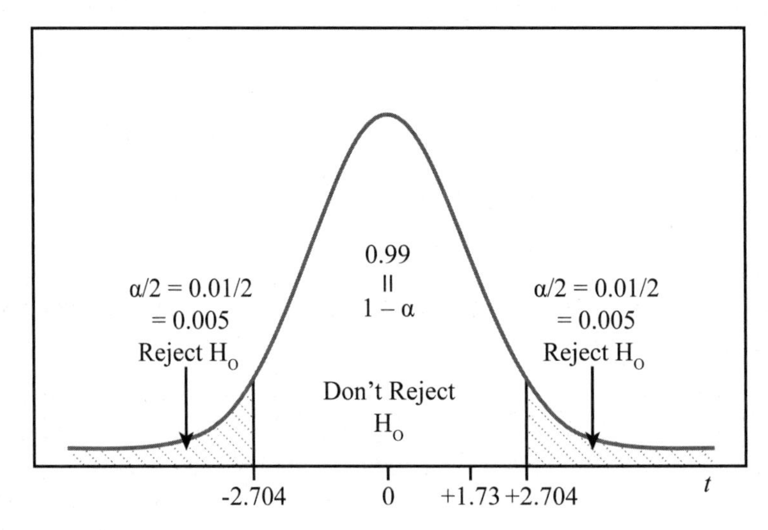

Figure 16.4

Hypothesis test for the registered nurses' salary example.

The calculated *t*-test statistic falls in the "Don't Reject H_0" region, so based on our result there is no significant difference between registered nurses' salaries in California and Massachusetts.

Some books use a simpler degrees of freedom equation in cases like this. They would use the smaller of $n_1 - 1$ and $n_2 - 1$. Using this, we would still get the same result for our example. At $\alpha = 0.01$ and degrees of freedom = 19, the critical *t*-value is ±2.861. Our calculated *t*-test statistic still falls in the "Don't Reject H_0" region and our decision won't change by using this critical value.

This procedure was based on the assumption that the standard deviations of the populations were unequal. What if this assumption is not true? I'm glad you asked!

Equal Population Variances When the Population Standard Deviations Are Unknown

This case differs from the previous one (unequal population variances) in two ways: the standard error of the difference between two means and the calculated *t*-test statistics.

We are assuming that $\sigma_1 = \sigma_2$, but that the values of σ_1 and σ_2 are unknown. Under these conditions, we calculate a *pooled variance* by combining two sample variances into one using the following equation:

$$s_p^2 = \frac{(n_1 - 1)s_1^2 + (n_2 - 1)s_2^2}{n_1 + n_2 - 2}$$

The *pooled standard deviation* is $s_p = \sqrt{s_p^2}$. The standard error of the difference between two means is as follows:

$$\hat{\sigma}_{\bar{x}_1 - \bar{x}_2} = s_p\sqrt{\frac{1}{n_1} + \frac{1}{n_2}}$$

The calculated *t*-test statistic in this case is:

$$t = \frac{(\bar{x}_1 - \bar{x}_2) - (\mu_1 - \mu_2)}{s_p\sqrt{\frac{1}{n_1} + \frac{1}{n_2}}} \text{ with } d.f = n_1 + n_2 - 2$$

Don't panic just yet. These equations look a whole lot better with numbers plugged in, so let's apply them to an example.

Bob has a very mysterious occurrence in his household—batteries seem to vanish into thin air. So Bob started buying them in 24-packs at the warehouse store, naively thinking that "these will last a long time." Wrong again—the more he bought, the faster they disappeared. Maybe it has something to do with certain teenagers listening to music on their portable CD players at a "brain-numbing" volume into the wee hours of the morning. Just a thought. Bob heard about the new "longer-lasting battery," and he wanted to investigate it. Let's say a company is promoting one of these batteries, claiming that its life is significantly longer than regular batteries. The hypothesis statement would be:

$$H_0 : \mu_1 - \mu_2 \leq 0$$

$$H_1 : \mu_1 - \mu_2 > 0$$

where:

μ_1 = the mean life of the long-lasting batteries

μ_2 = the mean life of the regular batteries

We'll choose to test this hypothesis at the $\alpha = 0.01$ level. The following data was collected measuring the battery life in hours for both types of batteries:

Raw Data for Battery Example

Long-Lasting Battery (Population 1):
51 44 58 36 48 53 57 40 49 44 60 50

Regular Battery (Population 2):
42 29 51 38 39 44 35 40 48 45

Using Excel, we can summarize this data in the following table.

Summarized Battery Data

Population in Hours	Sample Mean ($\bar{x}$)	Sample Standard Deviation (s)	Sample Size (n)
Long-lasting (1)	49.2	7.31	12
Regular (2)	41.1	6.40	10

In this example, since we have small samples, we'll have to make the assumption that the battery life in both populations is normally distributed in order to conduct this test. Now we can plug these numbers into the pooled standard deviation equation:

$$s_p = \sqrt{\frac{(n_1-1)s_1^2+(n_2-1)s_2^2}{n_1+n_2-2}} = \sqrt{\frac{(12-1)(7.31)^2+(10-1)(6.40)^2}{12+10-2}}$$

$$s_p = \sqrt{\frac{956.44}{20}} = 6.92 \text{ hours.}$$

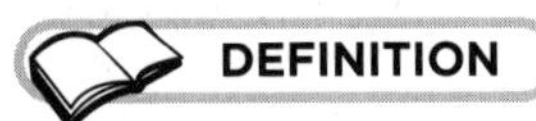

DEFINITION

The **pooled standard deviation** combines two sample variances into one variance and is calculated using $s_p = \sqrt{\frac{(n_1-1)s_1^2+(n_2-1)s_2^2}{n_1+n_2-2}}$.

We are now ready to determine our calculated t-test statistic as follows:

$$t = \frac{(\bar{x}_1-\bar{x}_2)-(\mu_1-\mu_2)}{s_p\sqrt{\frac{1}{n_1}+\frac{1}{n_2}}} = \frac{(49.2-41.1)-0}{(6.92)\sqrt{\frac{1}{12}+\frac{1}{10}}} = 2.73$$

The number of degrees of freedom for this test are:

$$d.f. = n_1 + n_2 - 2 = 12 + 10 - 2 = 20$$

The critical t-value, taken from Table 4 in Appendix B, for a one-tail (right) test using $\alpha = 0.01$ with $d.f. = 20$ is +2.528. This hypothesis test is shown graphically in Figure 16.5.

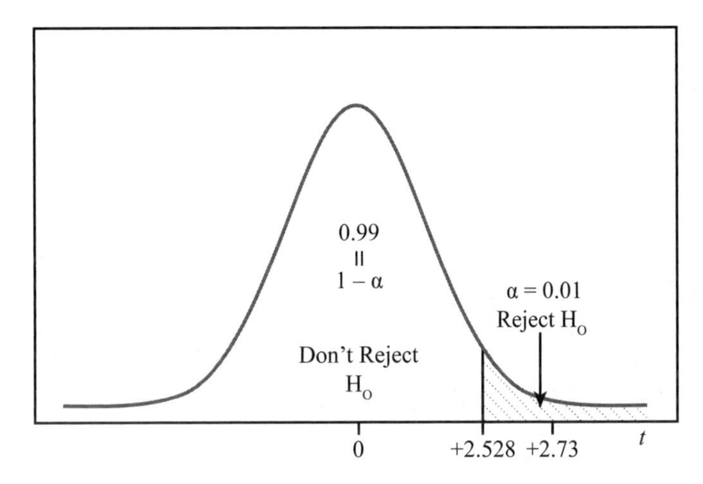

Figure 16.5

The hypothesis test for the battery life example.

According to Figure 16.5, our calculated t-test statistic of +2.73 is found in the "Reject H_0" region, which leads to our conclusion that the long-lasting batteries do indeed have a longer life than the regular batteries. Now that would have made Bob happy!

RANDOM THOUGHTS

These conditions are necessary for the hypothesis testing for differences between two means:

- The samples are independent of each other.
- If one or both of the sample sizes is small (<30), then the population must be normally distributed.
- If σ_1 and σ_2 are known, use the normal distribution to determine the rejection region(s).
- If σ_1 and σ_2 are unknown, approximate them with s_1 and s_2 and use the Student's t-distribution to determine the rejection region(s).

To Pool or Not to Pool?

In the previous section, we explored two cases for hypotheses testing between two populations: when the population variances are assumed to be equal and when they are unequal. I know you are asking, "When should I use each one?" Great question. Read on!

If the two samples are taken from the same population, then it is safe to assume equal population variances. For example, we could assume equal population variances if a researcher is interested in testing the effectiveness of a medication. The researcher chooses two different random samples from the same population: one sample receives the medication and the other sample receives the placebo. In this case, since the two samples are chosen from the same population, it is safe to assume equal population variances.

Another method is to compare the two sample variances, s_1^2 and s_2^2. These sample variances are estimates of the population variances. If they are close in value, then assume equal population variances. If they are not close, then assume unequal population variances. "How close is close?" you are asking. A simple rule of thumb is that if one of the sample variances is more than twice as large as the other sample variance, then it is safe to assume unequal population variances.

In addition, there are tests that you can perform to test for equal variances. However, these are outside the scope of this book.

Testing for Differences Between Means with Dependent Samples

Up to this point, all the samples that we have used in the chapter have been *independent samples.* Independent samples are not related in any way with each other. This is in contrast to *dependent samples*, where each observation of one sample is related to an observation in another sample.

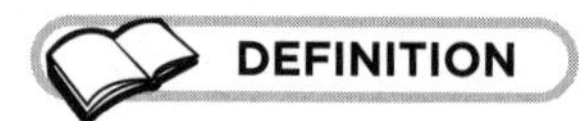

With **independent samples**, there is no relationship in the observations between the samples. With **dependent samples**, the observation from one sample is related to an observation from another sample.

An example of a dependent sample would be a weight-loss study. Each person is weighed at the beginning (Population 1) and end (Population 2) of the program. The change in weight of each person is calculated by subtracting the Population 2 weights from the Population 1 weights. Each observation from Population 1 is matched to an observation in Population 2. This hypothesis test is also known as a "matched-pair test." Dependent samples are tested differently than independent samples.

To demonstrate testing dependent samples, let's use the following example. A pharmaceutical company is introducing a new medication to lower cholesterol levels for patients. To test the effectiveness of their new medication, they selected a random sample of 9 people and measured their cholesterol levels before and after they took the medication. Cholesterol level is measured in milligrams per deciliter (mg/dL). The following table shows the results. The letter "*d*" refers to the difference between the cholesterol levels before and after taking the medication. The variable *d* is also called the matched-pair difference and is calculated as $d = x_1 - x_2$.

Differences in Cholesterol Levels Example

Patient	**1**	**2**	**3**	**4**	**5**	**6**	**7**	**8**	**9**
Before (x_1)	183	277	270	312	296	183	215	309	274
After (x_2)	189	231	220	266	293	163	227	285	254
d	-6	46	50	46	3	20	-12	24	20
d^2	36	2116	2500	2116	9	400	144	576	400

For future calculations, we will need:

$$\sum d = -6 + 46 + 50 + 46 + 3 + 20 - 12 + 24 + 20 = 191$$

$$\sum d^2 = 36 + 2{,}116 + 2{,}500 + 2{,}116 + 9 + 400 + 144 + 576 + 400 = 8{,}297$$

Patients' cholesterol levels before taking the medication will be considered Population 1, and after taking the medication will be considered Population 2. Because we are using the same patient's cholesterol level before and after taking the medication, these two samples are considered dependent.

Since the claim we are testing is $\mu_1 > \mu_2$, we can write our null and alternative hypotheses as:

$$H_0 : \mu_1 - \mu_2 \leq 0$$

$$H_1 : \mu_1 - \mu_2 > 0$$

where:

μ_1 = the average cholesterol level for patients before taking the new medication.

μ_2 = the average cholesterol level for patients after taking the new medication.

However, because we are only interested in the difference between the two populations, we can rewrite this statement as a single sample hypothesis as follows:

$$H_0 : \mu_d \leq 0$$

$$H_1 : \mu_d > 0$$

where μ_d is the mean of the difference between the two populations.

We will test this hypothesis using $\alpha = 0.05$.

Our next step is to calculate the mean, $\bar{d}$, and the standard deviation, s_d, of the matched-pair difference as follows:

$$\bar{d} = \frac{\sum_{i=1}^{n} d_i}{n} = \frac{191}{9} = 21.22 \text{ mg/dL}$$

This number tells us that the new medication lowers the cholesterol level by an average of about 21 mg/dL.

$$s_d = \sqrt{\frac{\sum_{i=1}^{n} d_i^2 - n(\bar{d})^2}{n-1}} = \sqrt{\frac{8{,}297 - 9(21.22)^2}{9-1}} = \sqrt{\frac{4{,}244.40}{8}} = 23.03 \text{ mg/dL}$$

TEST YOUR KNOWLEDGE

Does the equation for s_d look familiar? Yes, it is the same standard deviation equation you learned in Chapter 5.

If both populations follow the normal distribution, we use the Student's *t*-distribution because both sample sizes are less than 30 and σ_1 and σ_2 are unknown. The calculated *t*-test statistic is:

$$t = \frac{\bar{d} - \mu_d}{\frac{s_d}{\sqrt{n}}} = \frac{21.22 - 0}{\frac{23.03}{\sqrt{9}}} = 2.76$$

The number of degrees of freedom for this test are:

$$d.f. = n - 1 = 9 - 1 = 8$$

The critical *t*-value, taken from Table 4 in Appendix B, for a one-tail (right) test using $\alpha = 0.05$ with $d.f. = 8$ is +1.86. This hypothesis test is shown graphically in Figure 16.6.

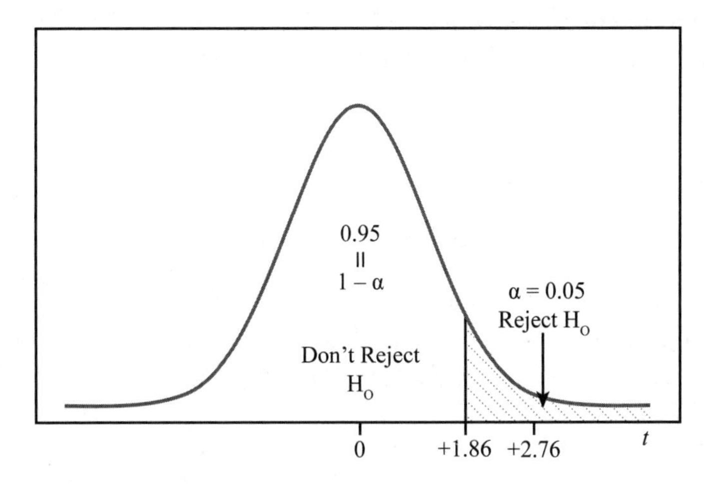

Figure 16.6
The hypothesis test for the cholesterol difference example.

According to Figure 16.6, our calculated *t*-test statistic of +2.76 falls into the "Reject H_0" region, which leads us to conclude that the new medication lowers cholesterol levels. Good news for the pharmaceutical company!

Letting Excel Do the Grunt Work

Excel performs many of the hypothesis tests that we've discussed in this chapter. Let me show you a few examples using this nifty tool. (First make sure that your Data Analysis Add-In is installed. Refer to the section "Installing the Data Analysis Add-In" from Chapter 2 if you don't see the command.) I'll start with the previous battery example. Follow these steps:

1. Open a blank Excel sheet and enter the data from the battery example in Columns A and B as shown in Figure 16.7.
2. From the Data tab at the top of the Excel window, choose Data Analysis and select *t*-Test: Two-Sample: Assuming Unequal Variances.
3. Click OK.

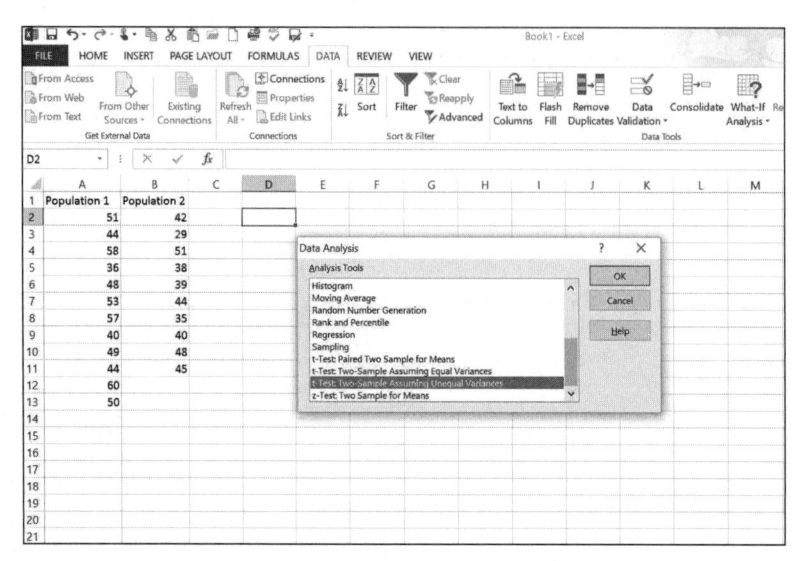

Figure 16.7

Data entry for the battery example.

4. In the *t*-Test: Two-Sample Assuming Unequal Variances dialog box, choose cells A1:A13 for the Variable 1 Range and cells B1:B11 for the Variable 2 Range. Set the Hypothesized Mean Difference to 0, Alpha to 0.01, Output Range to cell D2, and check the Labels box since we have the labels in the first cells, as shown in Figure 16.8.

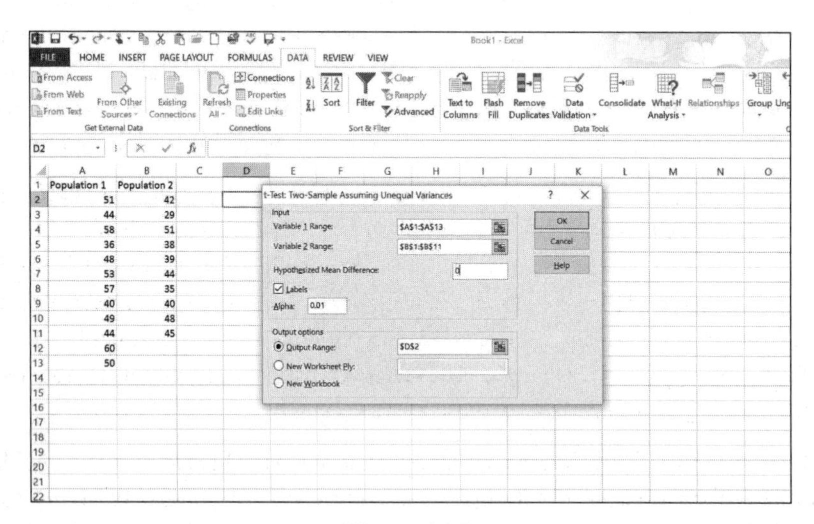

Figure 16.8

The t-Test: Two Sample Assuming Unequal Variances dialog box.

5. Click OK. The *t*-test output is shown in Figure 16.9.

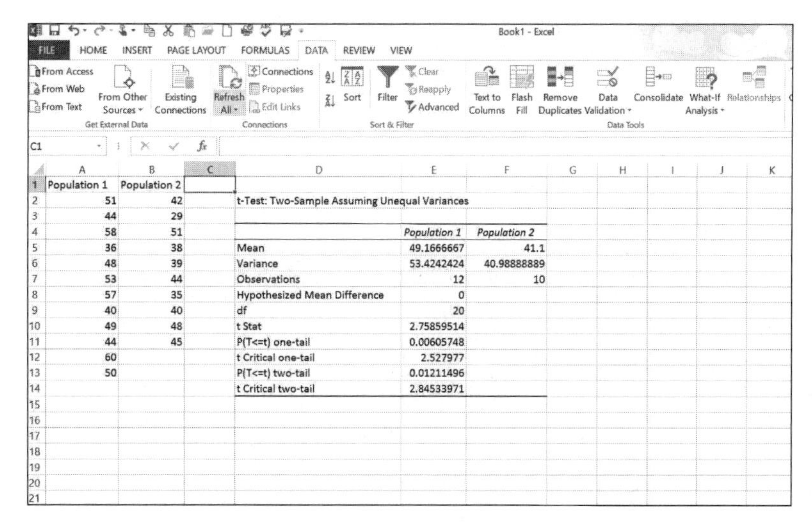

Figure 16.9

t-Test: Two Sample Assuming Unequal Variances output.

According to Figure 16.9, the calculated t-test statistic is 2.759, found in cell E10. This is the same value we calculated using the formula in the previous sections. The p-value of 0.006 is found in cell E11. Because the $p\text{-value} < \alpha$, we reject the null hypothesis.

If you look at the Excel Data Analysis tool box in Figure 16.7, you see that we can do hypothesis tests for all four cases we had in this chapter. You can use the z-Test: Two Samples for Means to do hypothesis testing when the population standard deviations (σ_1 and σ_2) are known. You can use t-Test: Two-Sample: Assuming Equal Variances to do hypothesis testing when the population standard deviations (σ_1 and σ_2) are unknown and assuming equal population variances. You can use t-Test: Paired Two Samples for Means to do hypothesis testing with dependent samples.

Let's see how to use Excel to do hypothesis testing when the population standard deviations (σ_1 and σ_2) are known, a z-test. We are going to use our previous example of SAT scores between North Carolina and South Carolina, where σ_1 for North Carolina = 200 and σ_2 for South Carolina = 150. I'm going to use a smaller sample to make the calculations easier and faster—we'll use 35 and 30 respectively, instead of 75 and 70 as we did in our example above using the formula. Just like the previous example, we will enter the data in columns A and B and choose z-Test: Two Sample for Means from the Data Analysis box.

In the z-Test: Two Sample for Means dialog box, choose cells A1:A36 for the Variable 1 Range and cells B1:B31 for the Variable 2 Range. Set the Hypothesized Mean Difference to 0, Alpha to 0.05, Variable 1 Variance to 40000 (which is $(200)^2$), Variable 2 Variance to 22500 (which is $(150)^2$), Output Range to cell D1, and check the Labels box since we have the labels in the first cells as shown in Figure 16.10.

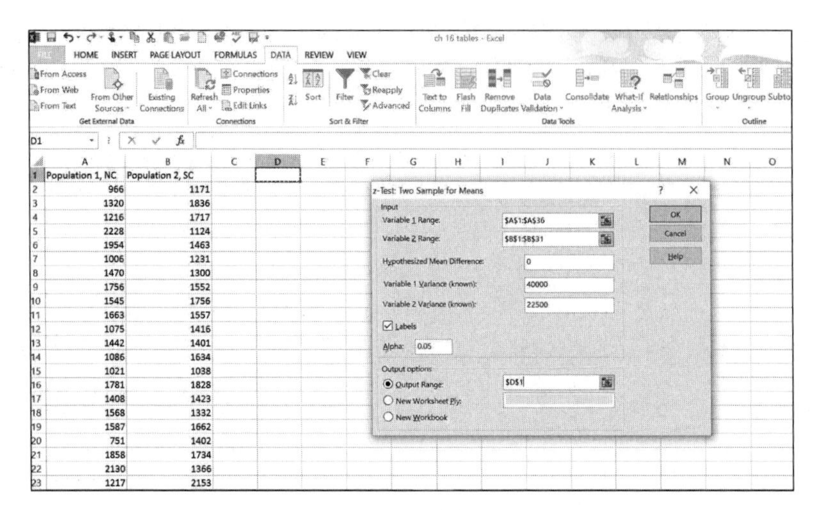

Figure 16.10

The z-Test: Two Sample for Means dialog box.

Click OK. The *z*-test output is shown in Figure 16.11.

ch 16 tables - Excel

	A	B	C	D	E	F
1	Population 1, NC	Population 2, SC		z-Test: Two Sample for Means		
2	966	1171				
3	1320	1836			*Population 1, NC*	*Population 2, SC*
4	1216	1717		Mean	1437.683084	1494.183578
5	2228	1124		Known Variance	40000	22500
6	1954	1463		Observations	35	30
7	1006	1231		Hypothesized Mean	0	
8	1470	1300		z	-1.298653607	
9	1756	1552		P(Z<=z) one-tail	0.097031416	
10	1545	1756		z Critical one-tail	1.644853627	
11	1663	1557		P(Z<=z) two-tail	0.194062832	
12	1075	1416		z Critical two-tail	1.959963985	
13	1442	1401				
14	1086	1634				
15	1021	1038				
16	1781	1828				
17	1408	1423				
18	1568	1332				
19	1587	1662				
20	751	1402				
21	1858	1734				
22	2130	1366				

Figure 16.11

The z-Test: Two Sample for Means output.

According to Figure 16.11, the calculated z-test statistic is -1.2987, found in cell E8, and the z-critical value is ±1.96, found in cell E12. The p-value is 0.194, found in cell E11. Because p-value > α, we don't reject the null hypothesis.

Let's go through one more example of how to use Excel to do hypothesis testing with dependent samples. We are going to use our previous example of the differences in cholesterol levels. Just like in the last Excel example, we will enter the data in columns A and B and choose t-Test: Paired Two Samples for Means from the Data Analysis box.

In the t-Test: Paired Two Sample for Means dialog box, choose cells A1:A10 for the Variable 1 Range and cells B1:B10 for the Variable 2 Range. Set the Hypothesized Mean Difference to 0, Alpha to 0.05, Output Range to cell D2, and check the Labels box since we have the labels in the first cells, as shown in Figure 16.12.

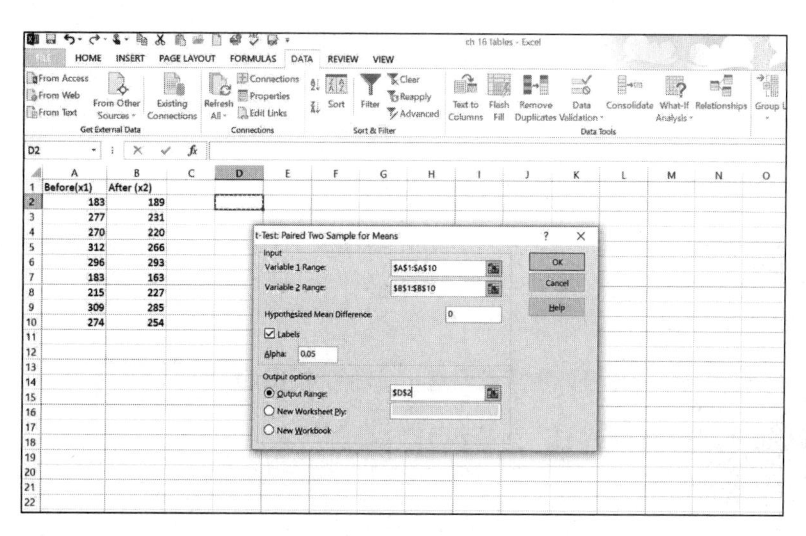

Figure 16.12

The t-Test: Paired Two Sample for Means dialog box.

Click OK. The t-test output is shown in Figure 16.13.

	A	B	C	D	E	F
1	Before(x1)	After (x2)				
2	183	189		t-Test: Paired Two Sample for Means		
3	277	231				
4	270	220			*Before(x1)*	*After (x2)*
5	312	266		Mean	257.66667	236.4444444
6	296	293		Variance	2602.5	1846.527778
7	183	163		Observations	9	9
8	215	227		Pearson Correlation	0.8937701	
9	309	285		Hypothesized Mean Differenc	0	
10	274	254		df	8	
11				t Stat	2.7643445	
12				P(T<=t) one-tail	0.0122549	
13				t Critical one-tail	1.859548	
14				P(T<=t) two-tail	0.0245098	
15				t Critical two-tail	2.3060041	

Figure 16.13

The t-Test: Paired Two Sample for Means output.

According to Figure 16.13, the calculated t-test statistic is 2.76, found in cell E11. The p-value is 0.012, found in cell E12. Because p-value $< \alpha$, we reject the null hypothesis.

Testing for Differences Between Proportions with Independent Samples

We can perform hypothesis testing to examine the difference between proportions of two populations as long as the sample size is large enough. Recall from Chapter 13, proportion data follow the binomial distribution, which can be approximated by the normal distribution under the following conditions.

$$np \geq 5 \text{ and } nq \geq 5$$

where:

p = the probability of a success in the population

q = the probability of a failure in the population ($q = 1 - p$)

Let's say that I want to test whether the proportion of males and females who use Facebook every day are the same. My hypotheses would be stated as:

$$H_0 : p_1 = p_2$$

$$H_1 : p_1 \neq p_2$$

where:

p_1 = the proportion of males who use Facebook every day

p_2 = the proportion of females who use Facebook every day

The following table summarizes the data from the Facebook samples:

Summarized Data for Facebook Samples

Population	Number of Successes (x)	Sample Size (n)
Male	207	300
Female	266	350

What can we conclude at $\alpha = 0.10$ level?

Our sample proportion of male Facebook users, $\bar{p}_1$, and female users, $\bar{p}_2$, can be found by:

$$\bar{p}_1 = \frac{x_1}{n_1} = \frac{207}{300} = 0.69 \text{ and } \bar{p}_2 = \frac{x_2}{n_2} = \frac{266}{350} = 0.76$$

To determine the calculated z-test statistic, we need to know the standard error of the difference between two proportions (that's a mouthful), $\sigma_{\bar{p}_1 - \bar{p}_2}$, which is found using:

$$\sigma_{\bar{p}_1 - \bar{p}_2} = \sqrt{\frac{p_1(1-p_1)}{n_1} + \frac{p_2(1-p_2)}{n_2}}$$

Our problem is that we don't know the values of p_1 and p_2, the actual population proportions of male and female Facebook users. The next best thing is to calculate the estimated standard error of the difference between two proportions, $\hat{\sigma}_{\bar{p}_1 - \bar{p}_2}$, using the following equation:

$$\hat{\sigma}_{\bar{p}_1 - \bar{p}_2} = \sqrt{(\hat{p})(1-\hat{p})\left(\frac{1}{n_1} + \frac{1}{n_2}\right)}$$

where $\hat{p}$, the estimated overall proportion of two populations, is found using the following equation:

$$\hat{p} = \frac{x_1 + x_2}{n_1 + n_2} = \frac{207+266}{300+350} = 0.728$$

For our Facebook example, the estimated standard error of the difference between two proportions is:

$$\hat{\sigma}_{\bar{p}_1-\bar{p}_2} = \sqrt{(0.728)(1-0.728)\left(\frac{1}{300}+\frac{1}{350}\right)} = 0.035$$

Now we can finally determine the calculated z-test statistic using:

$$z = \frac{(\bar{p}_1-\bar{p}_2)-(p_1-p_2)_{H_0}}{\hat{\sigma}_{\bar{p}_1-\bar{p}_2}}$$

For the Facebook example, our calculated z-test statistic becomes:

$$z = \frac{(\bar{p}_1-\bar{p}_2)-(p_1-p_2)_{H_0}}{\hat{\sigma}_{\bar{p}_1-\bar{p}_2}} = \frac{(0.69-0.76)-0}{0.035} = -2.00$$

BOB'S BASICS

The term $(p_1 - p_2)_{H_0}$ refers to the hypothesized difference between the two population proportions. When the null hypothesis is testing that there is no difference between population proportions, then the term $(p_1 - p_2)_{H_0}$ is set to 0.

The critical z-values for a two-tail test with $\alpha = 0.10$ are +1.65 and -1.65. Figure 16.14 shows this hypothesis test graphically.

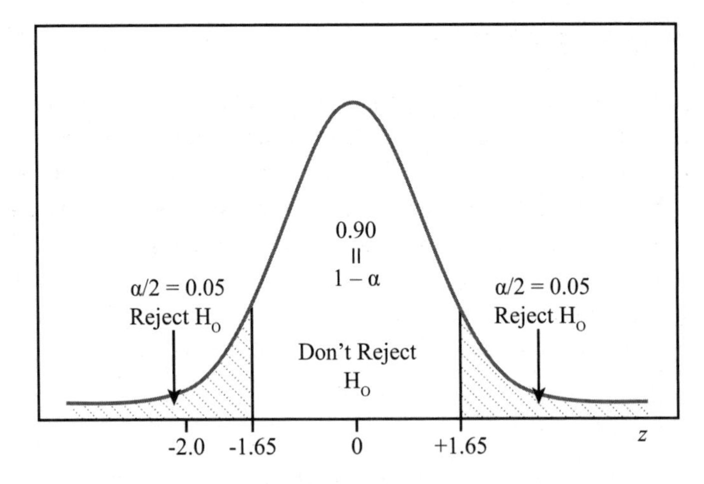

Figure 16.14
The hypothesis test for the Facebook example.

As you can see in Figure 16.14, the calculated z-test statistic of -2.00 falls in the "Reject H_0" region. Therefore, we conclude that the proportions of males and females who use Facebook every day are not equal to each other. I'm sure you are not surprised by this conclusion!

BOB'S BASICS

The standard error of the difference between two proportions describes the variation in the difference between two sample proportions and is calculated using $\hat{\sigma}_{\bar{p}_1-\bar{p}_2}$. The estimated standard error of the difference between two proportions approximates the variation in the difference between two sample proportions and is calculated using $\hat{\sigma}_{\bar{p}_1-\bar{p}_2} = \sqrt{(\hat{p})(1-\hat{p})\left(\frac{1}{n_1}+\frac{1}{n_2}\right)}$. The estimated overall proportion of two populations is the weighted average of two sample proportions and is calculated using $\hat{p} = \frac{x_1+x_2}{n_1+n_2}$.

The p-value for these samples can be found using the normal z distribution table found in Appendix B as follows:

$$2 \times P(z < -2.00) = 2 \times (0.0228) = 0.0456$$

This also confirms that we reject H_0 because the p-value $< \alpha$.

This completes our invigorating journey through the land of hypothesis testing.

Practice Problems

1. Test the hypothesis that the average SAT math scores from students in Pennsylvania and Ohio are different. A sample of 45 students from Pennsylvania had an average score of 552, whereas a sample of 38 Ohio students had an average score of 530. Assume the population standard deviations for Pennsylvania and Ohio are 105 and 114, respectively. Test at the $\alpha = 0.05$ level. What is the p-value for these samples?

2. A company tracks satisfaction scores based on customer feedback from individual stores on a scale of 0 to 100. The following data represents the customer scores from Stores 1 and 2.

   Store 1:

   90 87 93 75 88 96 90 82 95 97 78

   Store 2:

   82 85 90 74 80 89 75 81 93 75

   Assume population standard deviations are equal but unknown and that the population is normally distributed. Using $\alpha = 0.10$, test the hypothesis that the customer scores in the two stores are different..

3. A new diet program claims that participants will lose more than 15 pounds after completion of the program. The following data represents the before and after weights of nine individuals who completed the program. Test the claim at the $\alpha = 0.05$ level.

   Before: 221 215 206 185 202 197 244 188 218

   After: 200 192 195 166 187 177 227 165 201

4. Test the hypothesis that the proportion of home ownership in the state of Florida exceeds the national proportion at the $\alpha = 0.01$ level using the following data.

Population	Number of Successes	Sample Size
Florida	272	400
Nation	390	600

   What is the *p*-value for these samples?

5. Test the hypothesis that the average hourly wage for City A is more than $0.50 per hour above the average hourly wage in City B using the following sample data:

City	Average Wage	Standard Deviation	Sample Size
A	$9.80	$2.25	60
B	$9.10	$2.70	80

Test at the $\alpha = 0.05$ level. What is the p-value for this test?

6. Test the hypothesis that the average number of days that a home is on the market in City A is different from City B using the following sample data:

City A	12	8	19	10	26	4	15	20	18	25	7	11
City B	15	31	14	5	18	20	10	7	25	20	27	

Assume population standard deviations are unequal and that the population is normally distributed. Test the hypothesis using $\alpha = 0.10$.

The Least You Need to Know

- We use the normal distribution for the hypothesis test for the difference between means when $n \geq 30$ for both samples.
- We use the normal distribution for the hypothesis test for the difference between means when σ_1 and σ_2 are known, $n < 30$ for either sample, and both populations are normally distributed.
- We use the Student's t-distribution for the hypothesis test for the difference between means when σ_1 and σ_2 are unknown, $n < 30$ for either sample, and both populations are normally distributed.
- With dependent samples, the observation from one sample is related to an observation from another sample. With independent samples, there is no relationship in the observations between the samples.

APPENDIX

A

Solutions to "Practice Problems"

Chapter 1

1. Inferential statistics, because it would not be feasible to survey every Asian American household in the country. These results would be based on a sample of the population and used to make an inference on the entire population.
2. Inferential statistics, because it would not be feasible to survey every household in the country. These results would be based on a sample of the population and used to make an inference on the entire population.
3. Descriptive statistics, because Hank Aaron's home run total is based on the entire population, which is every at-bat in his career.
4. Descriptive statistics, because the average SAT score would be based on the entire population, which is the incoming freshman class.
5. Inferential statistics, because it would not be feasible to survey every American in the country. These results would be based on a sample of the population and used to make an inference on the entire population.

Chapter 2

1. Interval data, because temperature in degrees Fahrenheit does not contain a true zero point.
2. Ratio data, because monthly rainfall does have a true zero point.
3. Ordinal data, because a Master's degree is a higher level of education than a Bachelor's or high school degree. However, we cannot claim that a Master's degree is two or three times higher than the others.
4. Nominal data, because we cannot place the categories in any type of order.
5. Ratio data, because age does have a true zero point.
6. Definitely nominal data, unless you want to get into an argument about which is the lesser gender!
7. Interval data, because the difference between years is meaningful but a true zero point does not exist.
8. Nominal data, because we are not prepared to name one political party superior to another.
9. Nominal data, because these are simply unordered categories.
10. Ordinal data, because we can specify that "Above Expectation" is higher on the performance scale than the other two, but we cannot comment on the differences between the categories.
11. Nominal data, because we cannot claim a person wearing the number "10" is any better than a person wearing the number "4."
12. Ordinal data, because we cannot comment on the difference in performance between students. The top two students may be very far apart grade-wise, whereas the second and third students could be very close.
13. Ratio data, because these exam scores have a true zero point.
14. Nominal data, because there is no order in the states' categories.

Chapter 3

1.

Exam Grade	Number of Students
56–60	2
61–65	1
66–70	2
71–75	6
76–80	3
81–85	8
86–90	5
91–95	3
96–100	6

2.

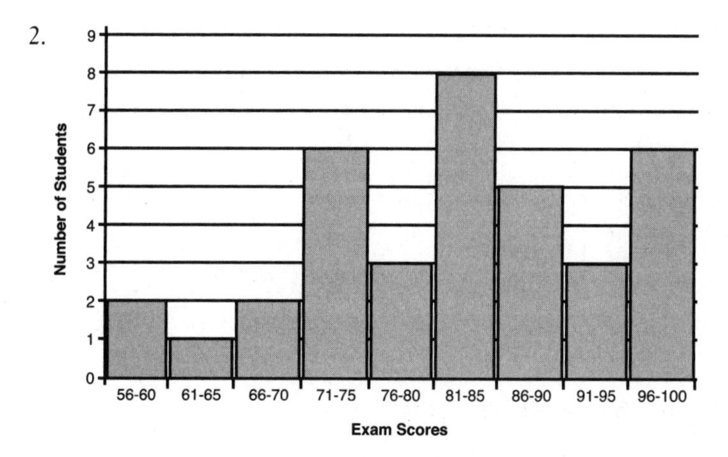

Histogram for Exam Grades.

3.

Exam Grade	Number of Students	Percentage	Cumulative Frequency
56–60	2	$^{2}/_{36}$ = 6%	2
61–65	1	$^{1}/_{36}$ = 3%	2 + 1 = 3
66–70	2	$^{2}/_{36}$ = 6%	3 + 2 = 5
71–75	6	$^{6}/_{36}$ = 17%	5 + 6 = 11
76–80	3	$^{3}/_{36}$ = 8%	11 + 3 = 14
81–85	8	$^{8}/_{36}$ = 22%	14 + 8 = 22
86–90	5	$^{5}/_{36}$ = 14%	22 + 5 = 27
91–95	3	$^{3}/_{36}$ = 8%	27 + 3 = 30
96–100	6	$^{6}/_{36}$ = 16%	30 + 6 = 36
	Total = 36		

4.

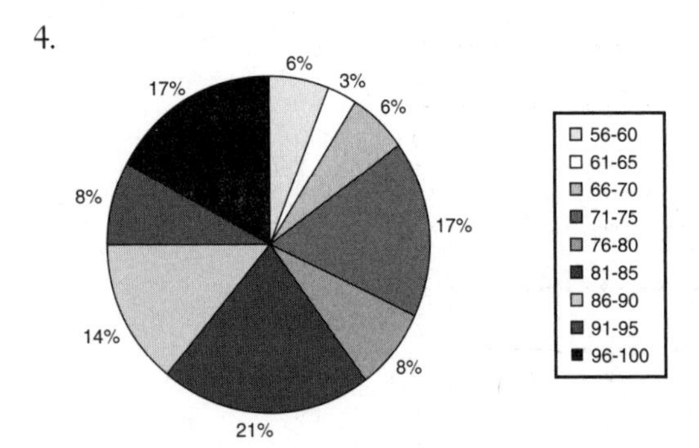

Pie Chart for Exam Grades

5.

5 | 8
6 | 0,2,6,8
7 | 2,2,2,4,5,5,8,9,9
8 | 1,1,2,3,4,5,5,5,6,6,6,8,9
9 | 1,2,5,6,8,8,9,9,9

Stem and Leaf for Problem 5

6.

5 (5) | 8
6 (0) | 0,2
6 (5) | 6,8
7 (0) | 2,2,2,4
7 (5) | 5,5,8,9,9
8 (0) | 1,1,2,3,4
8 (5) | 5,5,5,6,6,6,8,9
9 (0) | 1,2
9 (5) | 5,6,8,8,9,9,9

Stem and Leaf for Problem 6

Chapter 4

1. Mean = 15.9, Median = 17, Mode = 24
2. Mean = 81.7, Median = 82, Mode = 82
3. Mean = 32.7, Median = 32.5, Mode = 36 and 27
4. Mean = 7.2, Median = 6, Mode = 6
5. $\bar{x} = \frac{(22)(8)+(27)(37)+(32)(25)+(37)(48)+(42)(27)+(47)(10)}{8+37+25+48+27+10} = 34.5$ years
6. $\bar{x} = \frac{(3)(118)+(2)(125)+(1)(107)}{3+2+1} = 118.5$

Chapter 5

1.

x_i	x^2
20	400
15	225
24	576
10	100
8	64
19	361
24	576
$\sum_{i=1}^{n} x_i = 120$	$\sum_{i=1}^{n} x_i^2 = 2,302$

$\left(\sum_{i=1}^{n} x_i\right)^2 = (120)^2 = 14,400$, $s^2 = \frac{\sum_{i=1}^{n} x_i^2 - \frac{\left(\sum_{i=1}^{n} x_i\right)^2}{n}}{n-1} = \frac{2,302 - \frac{14,400}{7}}{6} = 40.8$, $s = \sqrt{40.8} = 6.4$,

Range = 24 – 8 = 16

2.

x_i	x^2
84	7,056
82	6,724
90	8,100
77	5,929
75	5,625
77	5,929
82	6,724
86	7,396
82	6,724
$\sum_{i=1}^{N} x_i = 735$	$\sum_{i=1}^{N} x_i^2 = 60{,}207$

$\left(\sum_{i=1}^{N} x_i\right)^2 = (735)^2 = 540{,}225$, $\sigma^2 = \dfrac{\sum_{i=1}^{N} x_i^2 - \dfrac{\left(\sum_{i=1}^{N} x_i\right)^2}{N}}{N} = \dfrac{60{,}207 - \dfrac{540{,}225}{9}}{9} = 20.2$ $\sigma = \sqrt{20.2} = 4.5$,

Range = 90 – 75 = 15

3. Range = 25, Variance = 75.4, Standard Deviation = 8.7

4. 2 5 6 6 <u>6</u> <u>8</u> 10 11 11 15

$Q1 = 6$ $Q2 = 7$ $Q3 = 11$

Note that the median of the data set is underlined.

5.

Midpoint (M)	Frequency (f)	fM	M^2	fM^2
22	8	176	484	3,872
27	37	999	729	26,973
32	25	800	1,024	25,600
37	48	1,776	1,369	65,712
42	27	1,134	1,764	47,628
47	10	470	2,209	22,090
	$\sum f = 155$	$\sum fM = 5{,}355$		$\sum fM^2 = 191{,}875$

$$\overline{x} = \frac{\sum_{i=1}^{k} f_i M_i}{\sum_{i=1}^{k} f_i} = \frac{5{,}355}{155} = 34.55 \text{ years}$$

$$s^2 = \frac{\sum_{i=1}^{k} fM^2 - n\overline{x}^2}{n-1} = \frac{191{,}875-(155)(34.55)^2}{155-1} = \frac{6{,}851.11}{154} = 44.49$$

$$s = \sqrt{44.49} = 6.67 \text{ years}$$

Chapter 6

1a. Empirical, because we have historical data for Derek Jeter's batting average.

1b. Classical, because we know the number of cards and the number of aces in the deck.

1c. If I have data from my last several rounds of golf, this would be empirical, otherwise subjective.

1d. Classical, because we can calculate the probability based on the lottery rules.

1e. Subjective, because I would not be collecting data for this experiment.

2a. Yes.

2b. No, probability cannot be greater than 1.

2c. No, probability cannot be greater than 100 percent.

2d. No, probability cannot be negative.

2e. Yes.

2f. Yes.

Chapter 7

1.a. $P(A) = \frac{177}{260} = 0.68$

1.b. $P(B) = \frac{152}{260} = 0.58$

1.c. $P(A') = \frac{83}{260} = 0.32$

1.d. $P(B') = \frac{108}{260} = 0.42$

1.e. $P(A \mid B) = \frac{98}{152} = 0.64$

1.f $P(A' \mid B) = \frac{54}{152} = 0.36$

1.g. $P(A \mid B') = \frac{79}{108} = 0.73$

1.h. $P(A \text{ and } B) = \frac{98}{260} = 0.38$

1.i. $P(A \text{ and } B') = \frac{79}{260} = 0.30$

1.j. P(A or B) = P(A) + P(B) – P(A and B) = 0.68 + 0.58 – 0.38 = 0.88

1.k. P(A or B') = P(A) + P(B′) – P(A and B’) = 0.68 + 0.42 – 0.30 = 0.80

1.l. $P(B \mid A) = \frac{P(B)P(A|B)}{P(B)P(A|B)+P(B')P(A|B')}$

$P(B \mid A) = \frac{(0.58)(0.64)}{(0.58)(0.64)+(0.42)(0.73)} = \frac{0.37}{0.37+0.31} = 0.54$

2.a. $P(A) = \frac{52}{125} = 0.42$

2.b. $P(B) = \frac{41}{125} = 0.33$

2.c. $P(A \text{ and } B) = \frac{23}{125} = 0.18$

2.d. P(A or B) = P(A) + P(B) – (A and B)

P(A or B) = 0.42 + 0.33 – 0.18 = 0.57

3. $3 \times 8 \times 4 \times 3 = 288$ different meals

4. There are $4 \times 4 \times 4 \times 4 \times 4 \times 4 \times 4 \times 4 \times 4 \times 4 = 1{,}048{,}576$ different ways to answer the exam. If only one of these sequences is correct, the probability is 1/1048576 = 0.00000095 that the student will correctly guess the correct sequence.

5. 13! = 6,227,020,800 different ordered arrangements

6. ${}_8P_3 = \frac{8!}{(8-3)!} = 8 \cdot 7 \cdot 6 = 336$

7. ${}_{10}P_2 = \frac{10!}{(10-2)!} = 10 \cdot 9 = 90$

8. ${}_{40}P_3 = \frac{40!}{(40-3)!} = 40 \cdot 39 \cdot 38 = 59{,}280$

9. $_{12}C_3 = \frac{12!}{3!(12-3)!} = \frac{12 \cdot 11 \cdot 10}{3 \cdot 2 \cdot 1} = 220$

10. $_{50}C_{12} = \frac{50!}{12!(50-12)!} = \frac{50 \cdot 49 \cdot 48 \cdot 47 \cdot 46 \cdot 45 \cdot 44 \cdot 43 \cdot 42 \cdot 41 \cdot 40 \cdot 39}{12 \cdot 11 \cdot 10 \cdot 9 \cdot 8 \cdot 7 \cdot 6 \cdot 5 \cdot 4 \cdot 3 \cdot 2 \cdot 1} = 121{,}399{,}651{,}100$

Chapter 8

1.

x_i	Probability	$x_i\,P(x_i)$	$(x_i - \mu)^2 P(x_i)$
0.1	0.3	$= (0.1)(0.3) = 0.03$	$= (0.1 - 0.145)^2(0.3) = 0.000608$
0.15	0.5	$= (0.15)(0.5) = 0.075$	$= (0.15 - 0.145)^2(0.5) = 0.000013$
0.2	0.2	$= (0.2)(0.2) = 0.04$	$= (0.2 - 0.145)^2(0.2) = 0.000605$

Mean (μ) = 0.03 + 0.075 + 0.04 = 0.145

Variance (σ^2) = 0.000608 + 0.000013 + 0.000605 = 0.001225

Standard deviation (σ) = $\sqrt{0.001225} = 0.035$

2.

Number of Cats (x_i)	Number of Families	Probability $P(x_i)$	x_i^2	$x_i^2 P(x_i)$
0	137	137/450 = 0.304	0	0
1	160	160/450 = 0.356	1	0.356
2	12	112/450 = 0.249	4	0.996
3	31	31/450 = 0.069	9	0.621
4	10	10/450 = 0.022	16	0.352
	Total = 350			$\sum_{i=1}^{n} x_i^2 P(x_i) = 2.325 = 2.325$

$$\mu = \sum_{i=1}^{n} x_i P(x_i) = (0)(0.304) + (1)(0.356) + (2)(0.249) + (3)(0.069) + (4)(0.022)$$

$\mu = 1.149$

$$\sigma^2 = \left(\sum_{i=1}^{n} x_i^2 P(x_i) \right) - \mu^2 = 2.325 - (1.149)^2 = 1.005$$

$$\sigma = \sqrt{\sigma^2} = \sqrt{1.005} = 1.002$$

Chapter 9

1. Because $n = 10$, $x = 7$, $p = 0.5$

   $P(7) = \frac{10!}{7!(10-7)!}(0.5)^7(0.5)^{10-7} = \left(\frac{10 \cdot 9 \cdot 8}{3 \cdot 2 \cdot 1}\right)(0.0078)(0.125)_V = 0.117$

2. Because $n = 6$, $x = 3$, $p = 0.75$

   $P(3) = \frac{6!}{3!(6-3)!}(0.75)^3(0.25)^{6-3} = \left(\frac{6 \cdot 5 \cdot 4}{3 \cdot 2 \cdot 1}\right)(0.4219)(0.0156) = 0.1316$

3. The probability of making at least 6 of 8 is P(6) + P(7) + P(8). Because $n = 8$, $p = 0.8$

   $P(6) = \frac{8!}{6!(8-6)!}(0.8)^6(0.2)^{8-6} = \left(\frac{8 \cdot 7}{2 \cdot 1}\right)(0.2621)(0.04) = 0.2936$

   $P(7) = \frac{8!}{7!(8-7)!}(0.8)^7(0.2)^{8-7} = (8)(0.2097)(0.2) = 0.3355$

   $P(8) = \frac{8!}{8!(8-8)!}(0.8)^8(0.2)^{8-8} = (1)(0.1678)(1) = 0.1678$

   Therefore, the probability of making at least 6 out of 8 is 0.2936 + 0.3355 + 0.1678 = 0.7969.

4. Because $n = 12$, $x = 6$, $p = 0.2$

   $P(6) = \frac{12!}{6!(12-6)!}(0.2)^6(0.8)^{12-6} = \left(\frac{12 \cdot 11 \cdot 10 \cdot 9 \cdot 8 \cdot 7}{6 \cdot 5 \cdot 4 \cdot 3 \cdot 2 \cdot 1}\right)(0.000064)(0.2621) = 0.0155$

5. The probability of no more than 2 out of the next 7 is P(0) + P(1) + P(2). Because $n = 7$, $p = 0.05$

   $P(0) = \frac{7!}{0!(7-0)!}(0.05)^0(0.95)^{7-0} = (1)(1)(0.6983) = 0.6983$

   $P(1) = \frac{7!}{1!(7-1)!}(0.05)^1(0.95)^{7-1} = (7)(0.05)(0.7351) = 0.2573$

   $P(2) = \frac{7!}{2!(7-2)!}(0.05)^2(0.95)^{7-2} = \left(\frac{7 \cdot 6}{2 \cdot 1}\right)(0.0025)(0.7738) = 0.0406$

   Therefore, the probability that no more than two of the next seven people will purchase is 0.6983 + 0.2573 + 0.0406 = 0.9962.

6. Since $n = 4$, $p = 0.335$

   $P(0) = \frac{4!}{0!(4-0)!}(0.335)^0(0.665)^{4-0} = (1)(1)(0.196) = 0.196$

   $P(1) = \frac{4!}{1!(4-1)!}(0.335)^1(0.665)^{4-1} = (4)(0.335)(0.294) = 0.394$

   $P(2) = \frac{4!}{2!(4-2)!}(0.335)^2(0.665)^{4-2} = (6)(0.112)(0.442) = 0.297$

   $P(3) = \frac{4!}{3!(4-3)!}(0.335)^3(0.665)^{4-3} = (4)(0.038)(0.665) = 0.101$

   $P(4) = \frac{4!}{4!(4-4)!}(0.335)^4(0.665)^{4-4} = (1)(0.013)(1) = 0.013$

7. $P(4) = \frac{10!}{4!(10-4)!}(0.6)^4(0.4)^{10-4} = \left(\frac{10 \cdot 9 \cdot 8 \cdot 7}{4 \cdot 3 \cdot 2 \cdot 1}\right)(0.1296)(0.004096) = 0.1115$

Chapter 10

1. $P(4) = \frac{(6)^4(2.71828)^{-6}}{4!} = \frac{(1{,}296)(0.002479)}{24} = 0.1339$

2. $P(5) = \frac{(7.5)^5(2.71828)^{-7.5}}{5!} = \frac{(23{,}730.469)(0.0005532)}{120} = 0.1094$

3. $P(x > 2) = 1 - P(x \le 2) = 1 - [P(x = 0) + P(x = 1) + P(x = 2)]$

   $P(0) = \frac{(4.2)^0(2.71828)^{-4.2}}{0!} = \frac{(1)(0.015)}{1} = 0.015$

   $P(1) = \frac{(4.2)^1(2.71828)^{-4.2}}{1!} = \frac{(4.2)(0.015)}{1} = 0.063$

   $P(2) = \frac{(4.2)^2(2.71828)^{-4.2}}{2!} = \frac{(17.64)(0.015)}{2} = 0.1323$

   $P(x > 2) = 1 - (0.015 + 0.063 + 0.1323) = 0.7897$

4. $P(x \le 3) = P(x = 0) + P(x = 1) + P(x = 2) + P(x = 3)$

   $P(0) = \frac{(3.6)^0(2.71828)^{-3.6}}{0!} = \frac{(1)(0.027324)}{1} = 0.0273$

   $P(1) = \frac{(3.6)^1(2.71828)^{-3.6}}{1!} = \frac{(3.6)(0.027324)}{1} = 0.0984$

   $P(2) = \frac{(3.6)^2(2.71828)^{-3.6}}{2!} = \frac{(12.96)(0.027324)}{2} = 0.1771$

   $P(3) = \frac{(3.6)^3(2.71828)^{-3.6}}{3!} = \frac{(46.656)(0.027324)}{6} = 0.2125$

   $P(x \le 3) = 0.0273 + 0.0984 + 0.1771 + 0.2125 = 0.5152$

5. $P(1) = \frac{(2.5)^1(2.71828)^{-2.5}}{1!} = \frac{(2.5)(0.082085)}{1} = 0.2052$

6. $P(x) = \frac{(np)^x(e)^{-(np)}}{x!}$, $n = 25, p = 0.05, np = 1.25$

   $P(2) = \frac{(1.25)^2(2.71828)^{-1.25}}{2!} = \frac{(1.5625)(0.286505)}{2} = 0.2238$

Chapter 11

1a. $z = \frac{x-\mu}{\sigma} = \frac{65.5-62.6}{3.7} = +0.78$ $P(z > 0.78) = 1 - P(z \le 0.78) = 1 - 0.7823 = 0.2177$

1b. $z = \frac{58.1-62.6}{3.7} = -1.22$, $P(z > -1.22) = 1 - P(z \le -1.22) = 1 - 0.1112 = 0.8888$

1c. $z_1 = \frac{70-62.6}{3.7} = +2.0$, $z_2 = \frac{61-62.6}{3.7} = -0.43$,

   $P(-0.43 \le z \le +2.0) = P(x \le 2) - P(z \le -0.43) = 0.9772 - 0.3336 = 0.6436$

2a. $z = \frac{x-\mu}{\sigma} = \frac{190{,}000-176{,}000}{22{,}300} = +0.63$, $P(z < 0.63) = 0.7357$

2b. $z = \frac{158{,}000-176{,}000}{22{,}300} = -0.81$, $P(z < -0.81) = 0.2090$

2c. $z_1 = \frac{168{,}000-176{,}000}{22{,}300} = -0.36$, $z_2 = \frac{150{,}000-176{,}000}{22{,}300} = -1.17$

   $P(-1.17 \le z \le -0.36) = P(z \le -0.36) - P(z \le -1.17) = 0.3594 - 0.1210 = 0.2384$

3a. $z = \frac{x-\mu}{\sigma} = \frac{31-37.5}{7.6} = -0.86$, P($z$ > -0.86) = 1 – P(z ≤ -0.86) = 1 – 0.1949 = 0.8051

3b. $z = \frac{42-37.5}{7.6} = +0.59$, P($z$ < 0.59) = 0.7224

3c. $z = \frac{45-37.5}{7.6} = +0.99$, $z = \frac{40-37.5}{7.6} = +0.33$

P(0.99 ≤ z ≤ 0.33) = P(z ≤ 0.99) – P(z ≤ 0.33) = 0.8389 – 0.6293 = 0.2096

4. For this problem, n = 14, p = 0.5, and q = 0.5. We can use the normal approximation since $np = nq$ = (14)(0.5) = 7. The binomial probabilities from the binomial table are P(x = 4, 5, or 6) = 0.0611 + 0.1222 + 0.1833 = 0.3666. Also, $\mu = np$ = (14)(0.5) = 7 and $\sigma = \sqrt{npq} = \sqrt{(14)(0.5)(0.5)} = 1.871$. The normal approximation would be finding P(3.5 ≤ x ≤ 6.5).

   $z_1 = \frac{x-\mu}{\sigma} = \frac{6.5-7}{1.871} = -0.27$, $z_1 = \frac{3.5-7}{1.871} = -1.87$

   P(-1.87 ≤ z ≤ -0.27) = P(z ≤ -0.27) – P(z ≤ -1.87) = 0.3936 – 0.0307 = 0.3629

5a. $z_1 = \frac{x-\mu}{\sigma} = \frac{97-92}{4} = 1.25$, P($z$ > 1.25) = 1 – P(z ≤ 1.25) = 1 – 0.8944 = 0.1056

5b. $z_1 = \frac{90-92}{4} = -0.5$ P(z > -0.5) = 1 – P(z ≤ -0.5) = 1 – 0.3085 = 0.6915

6a. $z_1 = \frac{x-\mu}{\sigma} = \frac{5{,}000-4{,}580}{550} = +0.76$ $z_2 = \frac{4{,}000-4{,}580}{550} = -1.05$

P(+ 0.76 ≤ z ≤ -1.05) = P(z ≤ 0.76) – P(z ≤ -1.05) = 0.7764 – 0.1469 = 0.6295

6b. $z = \frac{4{,}200-4{,}580}{550} = -0.69$, P($z$ < -0.69) = 0.2451

7. Using the empirical rule, 95.5 percent of the values will fall within 2 standard deviations from the mean.

   95.5 percent of the data will fall in $\mu \pm 2\sigma$

   $\mu + 2\sigma = 75 + 2(10) = 95$, $\mu - 2\sigma = 75 - 2(10) = 55$

   Therefore, 95.5 percent of the data values should fall between 55 and 95.

Chapter 12

1. $k = \frac{N}{n} = \frac{75{,}000}{500} = 150$
2. If every employee belonged to a particular department, certain departments could be chosen for the survey, with every individual in those departments asked to participate. Other answers are also possible.
3. If each employee can be classified as either a manager or a non-manager, ensure that the sample proportion for each type is similar to the proportion of managers and non-managers in the company. Other answers are also possible.

Chapter 13

1a. $\sigma_{\bar{x}} = \frac{\sigma}{\sqrt{n}} = \frac{10}{\sqrt{15}} = 2.58$

1b. $\sigma_{\bar{x}} = \frac{\sigma}{\sqrt{n}} = \frac{4.7}{\sqrt{12}} = 1.36$

1c. $\sigma_{\bar{x}} = \frac{\sigma}{\sqrt{n}} = \frac{7}{\sqrt{20}} = 1.57$

2a. $\sigma_{\bar{x}} = \frac{\sigma}{\sqrt{n}} = \frac{7.5}{\sqrt{9}} = 2.5$, $z = \frac{\bar{x}-\mu}{\sigma_{\bar{x}}} = \frac{17-16}{2.5} = +0.40$, $P(z \le 0.40) = 0.6554$

2b. $z = \frac{18-16}{2.5} = +0.80$, $P(z > 0.80) = 1 - P(z \le 0.80) = 1 - 0.7881 = 0.2119$

2c. $z_1 = \frac{16.5-16}{2.5} = +0.20$, $z_2 = \frac{14.5-16}{2.5} = -0.60$

$P(+0.20 \le z \le -0.60) = P(z \le 0.20) - P(z \le -0.60) = 0.5793 - 0.2743 = 0.3050$

3a. $\sigma_p = \sqrt{\frac{P(1-P)}{n}} = \sqrt{\frac{0.25(1-0.25)}{200}} = 0.0306$

3b. $\sigma_p = \sqrt{\frac{P(1-P)}{n}} = \sqrt{\frac{0.42(1-0.42)}{100}} = 0.0494$

3c. $\sigma_p = \sqrt{\frac{P(1-P)}{n}} = \sqrt{\frac{0.06(1-0.06)}{175}} = 0.0179$

4a. $\sigma_p = \sqrt{\frac{P(1-P)}{n}} = \sqrt{\frac{0.32(1-0.32)}{160}} = 0.0369$, $z = \frac{\bar{p}-p}{\sigma_p} = \frac{0.30-0.32}{0.0369} = -0.54$, $P(z \le -0.54) = 0.2946$

4b. $z = \frac{0.36-0.32}{0.0369} = +1.08$, $P(z > 1.08) = 1 - P(z \le 1.08) = 1 - 0.8599 = 0.1401$

4c. $z_1 = \frac{0.37-0.32}{0.0369} = +1.36$, $z_2 = \frac{0.29-0.32}{0.0369} = -0.81$

$P(+1.36 \le z \le -0.81) = P(z \le 1.36) - P(z \le -0.81) = 0.9131 - 0.2090 = 0.7041$

5. $p = \frac{24}{60} = 0.4$, $\bar{p} = \frac{30}{60} = 0.5$ and

$\sigma_p = \sqrt{\frac{P(1-P)}{n}} = \sqrt{\frac{0.4(1-0.4)}{60}} = 0.063$, $z = \frac{0.5-0.4}{0.063} = +1.51$,

$P(z \ge 1.51) = 1 - P(z < 1.51) = 1 - 0.9345 = 0.0655$

Chapter 14

1. $\sigma_{\bar{x}} = \frac{\sigma}{\sqrt{n}} = \frac{7.6}{\sqrt{40}} = 1.20$, $z_{\alpha/2} = 2.17$

Upper Limit = $\bar{x} + z_{\alpha/2}\sigma_{\bar{x}} = 31.3 + (2.17)(1.20) = 33.90$

Lower Limit = $\bar{x} - z_{\alpha/2}\sigma_{\bar{x}} = 31.3 - (2.17)(1.20) = 28.70$

2. $n = \left(\frac{z_{\alpha/2}\sigma}{ME}\right)^2 = \left(\frac{(2.33)(15)}{5}\right)^2 = 48.9 \approx 49$

3. This is a trick question! The sample size is too small to be used from a population that is not normally distributed. This question goes beyond the scope of this book. You would need to consult a statistician.

4. Using Excel, we can calculate $\bar{x} = 13.9$ and $s = 6.04$.

   $s_{\bar{x}} = \frac{s}{\sqrt{n}} = \frac{6.04}{\sqrt{30}} = 1.10$

   Upper Limit $= \bar{x} + z_{\alpha/2} s_{\bar{x}} = 13.9 + (1.65)(1.10) = 15.72$

   Lower Limit $= \bar{x} - z_{\alpha/2} s_{\bar{x}} = 13.9 - (1.65)(1.10) = 12.09$

5. Using Excel, we can calculate $\bar{x} = 46.92$.

   $\sigma = 12.7$, $z_{\alpha/2} = 1.88$, and $\sigma_{\bar{x}} = \frac{\sigma}{\sqrt{n}} = \frac{12.7}{\sqrt{12}} = 3.67$

   Upper Limit $= \bar{x} + z_{\alpha/2} \sigma_{\bar{x}} = 46.92 + (1.88)(3.67) = 53.82$

   Lower Limit $= \bar{x} - z_{\alpha/2} \sigma_{\bar{x}} = 46.92 - (1.88)(3.67) = 40.02$

6. Using Excel, we can calculate $\bar{x} = 119.64$, $s = 11.29$, $s_{\bar{x}} = \frac{s}{\sqrt{n}} = \frac{11.29}{\sqrt{11}} = 3.40$.

   For a 98 percent confidence interval with $n - 1 = 11 - 1 = 10$ degrees of freedom, $t_{\alpha/2} = 2.764$.

   Upper Limit $= \bar{x} + t_{\alpha/2} s_{\bar{x}} = 119.64 + (2.764)(3.40) = 129.04$

   Lower Limit $= \bar{x} - t_{\alpha/2} s_{\bar{x}} = 119.64 - (2.764)(3.40) = 110.24$

7. This is another trick question! The sample size is too small to be used from a population that is not normally distributed. This question goes beyond the scope of this book. You would need to consult a statistician.

8. $\bar{p} = \frac{11}{200} = 0.055$. Since $n\bar{p} = (200)(0.055) = 11$ and $n\bar{q} = (200)(0.945) = 189$, we can use the normal approximation.

   $\hat{\sigma}_p = \sqrt{\frac{\bar{p}(1-\bar{P})}{n}} = \sqrt{\frac{(0.055)(1-0.055)}{200}} = 0.0161$, $z_{\alpha/2} = 1.96$

   Upper Limit $= \bar{p} + z_{\alpha/2} \hat{\sigma}_p = 0.055 + (1.96)(0.0161) = 0.087$

   Lower Limit $= \bar{p} - z_{\alpha/2} \hat{\sigma}_p = 0.055 - (1.96)(0.0161) = 0.023$

9. $n = \frac{(z_{\alpha/2})^2 pq}{(ME)^2} = \frac{(2.05)^2(0.55)(0.45)}{(0.04)^2} = 650$

Chapter 15

1. $H_0 : \mu \geq 40$, $H_1 : \mu < 40$, $n = 50$, $\sigma = 12.5$ years, $\sigma_{\bar{x}} = \frac{\sigma}{\sqrt{n}} = \frac{12.5}{\sqrt{50}} = 1.768$ years, $\bar{x} = 38.7$ years $z_c = -1.65$

   $z = \frac{\bar{x}-\mu}{\sigma_{\bar{x}}} = \frac{38.7-40}{1.768} = -0.7354$

   Since $|z| < |z_c|$, we do not reject H_0 and conclude that we do not have enough evidence to support the claim that the average age is less than 40 years old.

2. $H_0: \mu \leq 1000$, $H_1: \mu > 1000$, $n = 32$, $\sigma = 325$ hours, $\bar{x} = 1190$ hours,
   $\sigma_{\bar{x}} = \frac{\sigma}{\sqrt{n}} = \frac{325}{\sqrt{32}} = 57.45$ hours, $z_c = \pm 2.05$
   $z = \frac{\bar{x}-\mu}{\sigma_{\bar{x}}} = \frac{1190-1000}{57.45} = 3.307$
   Since the calculated z-test statistic of 3.307 > the critical value of 2.05, we reject H_0 and conclude the average light bulb life exceeds 1,000 hours.

3. $H_0: \mu \geq 30$, $H_1: \mu < 30$, $n = 42$, $\sigma = 8$ minutes, $\sigma_{\bar{x}} = \frac{\sigma}{\sqrt{n}} = \frac{8.0}{\sqrt{42}} = 1.23$, $z_c = -2.33$
   $z = \frac{\bar{x}-\mu}{\sigma_{\bar{x}}} = \frac{26.9-30}{1.23} = -2.52$
   Since $|Z| > |Z_c|$, we reject H_0 and conclude that the average delivery time is less than 30 minutes.

4. $H_0: \mu = \$32{,}700$, $H_1: \mu \neq \$32{,}700$, $n = 40$, $\sigma = \$950$, $\sigma_{\bar{X}} = \frac{\sigma}{\sqrt{n}} = \frac{\$950}{\sqrt{40}} = \$150.20$, $z_c = \pm 1.96$
   $z = \frac{\bar{x}-\mu}{\sigma_{\bar{X}}} = \frac{32{,}450-32{,}700}{150.20} = -1.66$, p-value = (2)[P($z \leq$ -1.66)] = (2)(0.0485) = 0.0970
   Since p-value > α, we do not reject H_0 and conclude that we do not have enough evidence to contradict the claim that the average graduating college student has $32,700 in student loan debt.

5. $H_0: \mu = 1500$, $H_1: \mu \neq 1500$, $n = 70$, $\sigma = 310$, $\sigma_{\bar{X}} = \frac{\sigma}{\sqrt{n}} = \frac{310}{\sqrt{70}} = 37.05$, $z_c = \pm 1.65$,
   $z = \frac{\bar{x}-\mu}{\sigma_{\bar{X}}} = \frac{1435-1500}{37.05} = -1.75$, p-value = (2)[P($z \leq$ -1.75)] = (2)(0.0401) = 0.0802
   Since p-value < α, we reject H_0 and conclude that the average SAT score does not equal 1500.

6. $H_0: \mu \leq 35$, $H_1: \mu > 35$, $\bar{x} = 37.9$, $s = 6.74$, $n = 10$, $d.f. = n - 1 = 9$, $s_{\bar{x}} = \frac{s}{\sqrt{n}} = \frac{6.74}{\sqrt{10}} = 2.13$, $t_c = 2.398$,
   $t = \frac{\bar{x}-\mu}{s_{\bar{x}}} = \frac{37.9-35}{2.13} = 1.36$
   Since $t < t_c$, we do not reject H_0 and conclude that the average class size is not significantly greater than 35 students.

7. $H_0: \mu \leq 7$, $H_1: \mu > 7$, $\bar{x} = 8.2$, $s = 4.29$, $n = 30$, $s_{\bar{x}} = \frac{s}{\sqrt{n}} = \frac{4.29}{\sqrt{30}} = 0.78$, $z_c = +1.65$,
   $z = \frac{\bar{x}-\mu}{\sigma_{\bar{X}}} = \frac{8.2-7}{0.78} = +1.54$
   p-value = P(z > + 1.54) = 1 – P($z \leq$ + 1.54) = 1 – 0.9382 = 0.0618
   Since $z < z_c$ or p-value > α, we do not reject H_0 and conclude that average gasoline consumption in the United States does not exceed seven liters per car per day.

8. $H_0: p \geq 0.40$, $H_1: p < 0.40$, $\sigma_p = \sqrt{\frac{p(1-P)}{n}} = \sqrt{\frac{(0.4)(1-0.4)}{175}} = 0.037$, $z = \frac{\bar{p}-p}{\sigma_p} = \frac{0.30-0.4}{0.037} = -2.70$
   p-value = P($z \leq$ -2.70) = 0.035, z_c = -2.33
   Since p-value < α, we reject H_0 and conclude that the proportion of Republicans is less than 40 percent.

9. $H_0: p = 0.65, H_1: p \neq 0.65,\ \sigma_p = \sqrt{\frac{p(1-P)}{n}} = \sqrt{\frac{(0.65)(1-0.65)}{225}} = 0.032,\ z = \frac{\bar{p}-p}{\sigma_p} = \frac{0.69-0.65}{0.032} = +1.25$.

   p-value = (2)P(z > + 1.25) = (2)[1 – P($z \leq$ + 1.25)] = (2)(1 – 0.8944) = 0.2122

   Since p-value > α, we fail to reject H_0 and conclude that the proportion of teenagers who exceed their minutes is not different from 65 percent.

10. $H_0: \mu \geq 15, H_1: \mu < 15, n = 60, \sigma = 5,\ \sigma_{\bar{x}} = \frac{\sigma}{\sqrt{n}} = \frac{5}{\sqrt{60}} = 0.645,\ z = \frac{\bar{x}-\mu}{\sigma_{\bar{X}}} = \frac{13.5-15}{0.645} = -2.33$.

    p-value = P($z \leq$ -2.33) = 0.0099

    Since p-value < α, we reject H_0 and conclude that the average number of hours worked is less than 15.

Chapter 16

1. $H_0: \mu_1 = \mu_2, H_1: \mu_1 \neq \mu_2$, PA = 1, Ohio = 2

   $$\sigma_{\bar{x}_1-\bar{x}_2} = \sqrt{\frac{\sigma_1^2}{n_1}+\frac{\sigma_2^2}{n_2}} = \sqrt{\frac{(105)^2}{45}+\frac{(114)^2}{38}} = 24.22,\ z_c = \pm 1.96$$

   $$z = \frac{(\bar{x}_1-\bar{x}_2)-(\mu_1-\mu_2)}{\sigma_{\bar{x}1-\bar{x}2}} = \frac{(552-530)-0}{24.22} = +0.91$$

   Since $z < z_c$, we do not reject H_0 and conclude there is not enough evidence to support a difference between the two states.

   p-value = (2)P[z > +0.91)] = (2)[1 – P($z \leq$ +0.91)] = (2)(1 – 0.8186) = 0.3628

2. $H_0: \mu_1 = \mu_2, H_1: \mu_1 \neq \mu_2,\ \bar{x}_1 = 88.3,\ s_1 = 7.30,\ \bar{x}_2 = 82.4,\ s_2 = 6.74,$

   $$s_p = \sqrt{\frac{(n_1-1)s_1^2+(n_2-1)s_2^2}{n_1+n_2-2}} = \sqrt{\frac{(10)(7.30)^2+(9)(6.74)^2}{11+10-2}} = 7.04$$

   $$\hat{\sigma}_{\bar{x}_1-\bar{x}_2} = s_p\sqrt{\frac{1}{n_1}+\frac{1}{n_2}} = (7.04)\sqrt{\frac{1}{11}+\frac{1}{10}} = (7.04)\sqrt{0.1909} = 3.08$$

   $$t = \frac{(\bar{x}_1-\bar{x}_2)-(\mu_1-\mu_2)}{\hat{\sigma}_{\bar{x}_1-\bar{x}_2}} = \frac{(88.3-82.4)-0}{3.08} = +1.92$$

   $d.f. = n_1 + n_2 - 2 = 11 + 10 - 2 = 19,\ t_c = \pm 1.729$

   Since $t > t_c$, we reject H_0 and conclude the satisfaction scores are not equal between the two stores.

3. $H_0 : \mu_d \leq 15, H_1 : \mu_d > 15, \sum d = 21+23+11+19+15+20+17+23+17 = 166$

$$\sum d^2 = 441+529+121+361+225+400+289+529+289 = 3{,}184$$

$$\bar{d} = \frac{\Sigma d}{2} = \frac{166}{9} = 18.44\,,$$

$$s_d = \sqrt{\frac{\sum d^2 - \frac{(\sum d)^2}{n}}{n-1}} = \sqrt{\frac{3{,}184 - \frac{(166)^2}{9}}{9-1}} = \sqrt{\frac{122.22}{8}} = 3.91$$

$$t = \frac{\bar{d}-\mu_d}{\frac{s_d}{\sqrt{n}}} = \frac{18.44-15}{\frac{3.91}{\sqrt{9}}} = \frac{3.44}{1.30} = +2.64\,,\ d.f. = n-1 = 9-1 = 8,\ t_c = +1.860$$

Since $t > t_c$, we reject H_0 and conclude the weight loss program claim is valid.

4. $H_0 : p_1 \leq p_2, H_1 : p_1 > p_2$, Pop 1 = Florida, Pop 2 = Nation

$$\bar{p}_1 = \frac{x_1}{n_1} = \frac{272}{400} = 0.68\,,\ \bar{p}_2 = \frac{x_2}{n_2} = \frac{390}{600} = 0.65\,,\ \hat{p} = \frac{x_1+x_2}{n_1+n_2} = \frac{272+390}{400+600} = 0.662$$

$$\hat{\sigma}_{\bar{p}_1-\bar{p}_2} = \sqrt{(\hat{p})(1-\hat{p})\left(\frac{1}{n_1}+\frac{1}{n_2}\right)} = \sqrt{(0.662)(1-0.662)\left(\frac{1}{400}+\frac{1}{600}\right)} = 0.0305$$

$$z = \frac{(\bar{p}_1-\bar{p}_2)-(p_1-p_2)_{H_0}}{\hat{\sigma}_{\bar{p}_1-\bar{p}_2}} = \frac{(0.68-0.65)-0}{0.0305} = +0.98\,,\ z_c = +2.33$$

Since $z < z_c$, we do not reject H_0 and conclude there is not enough evidence to support the claim that the proportion of home ownership in Florida is greater than the national proportion.

p-value = $P(z > +0.98) = 1 - P(z \leq +0.98) = (1 - 0.8365) = 0.1635$

5. $H_0 : \mu_1 - \mu_2 \leq 0.50, H_1 : \mu_1 - \mu_2 > 0.50$, Pop 1 = City A, Pop 2 = City B.

$$\sigma_{\bar{x}_1-\bar{x}_2} = \sqrt{\frac{s_1^2}{n_1}+\frac{s_2^2}{n_2}} = \sqrt{\frac{(2.25)^2}{60}+\frac{(2.70)^2}{80}} = 0.419$$

$$Z = \frac{(\bar{x}_1-\bar{x}_2)-(\mu_1-\mu_2)}{\sqrt{\frac{s_1^2}{n_1}+\frac{s_2^2}{n_2}}} = \frac{(9.80-9.10)-(0.50)}{0.419} = +0.477$$

p-value = $P(z > +0.48) = 1 - P(z \leq +0.48) = 1 - 0.6844 = 0.3156$

Since p-value > α, we don't reject H_0 and conclude that the average hourly wage in City A is not more than \$0.50 higher than City B.

6. $H_0 : \mu_1 = \mu_2,\ H_1 : \mu_1 \neq \mu_2,\ \bar{x}_1 = 14.58,\ s_1 = 7.09,\ \bar{x}_2 = 17.45,\ s_2 = 8.26$

$$\sigma_{\bar{x}_1-\bar{x}_2} = \sqrt{\frac{s_1^2}{n_1}+\frac{s_2^2}{n_2}} = \sqrt{\frac{(7.09)^2}{12}+\frac{(8.26)^2}{11}} = 3.224,$$

$$\frac{s_1^2}{n_1} = \frac{(7.09)^2}{12} = 4.19,$$

$$\frac{s_2^2}{n_2} = \frac{(8.26)^2}{11} = 6.20$$

$$d.f. = \frac{\left(\frac{s_1^2}{n_1}+\frac{s_2^2}{n_2}\right)^2}{\frac{\left(\frac{s_1^2}{n_1}\right)^2}{n_1-1}+\frac{\left(\frac{s_2^2}{n_2}\right)^2}{n_2-1}} = \frac{(4.19+6.20)^2}{\frac{(4.19)^2}{12-1}+\frac{(6.20)^2}{11-1}} = \frac{107.98}{5.447} \approx 20$$

$$t = \frac{(\bar{x}_1-\bar{x}_2)-(\mu_1-\mu_2)}{\sigma_{\bar{x}_1-\bar{x}_2}} = \frac{(14.58-17.45)-0}{3.224} = -0.890,\ t_c = \pm 1.725$$

Since $|t| < |t_c|$, we fail to reject H_0 and conclude that the number of days that a home is on the market in City A does not differ from City B.

APPENDIX

B

Statistical Tables

Source: Mr. Carl Schwarz, www.stat.sfu.ca/~cschwarz/. Used with permission.

Binomial Probability Tables

The following table provides the probability of exactly x successes in n trials for various values of p.

Table 1 Binomial Probability Tables

Values of p

n	x	0.1	0.2	0.3	0.4	0.5	0.6	0.7	0.8	0.9
2	0	0.8100	0.6400	0.4900	0.3600	0.2500	0.1600	0.0900	0.0400	0.0100
	1	0.1800	0.3200	0.4200	0.4800	0.5000	0.4800	0.4200	0.3200	0.1800
	2	0.0100	0.0400	0.0900	0.1600	0.2500	0.3600	0.4900	0.6400	0.8100
3	0	0.7290	0.5120	0.3430	0.2160	0.1250	0.0640	0.0270	0.0080	0.0010
	1	0.2430	0.3840	0.4410	0.4320	0.3750	0.2880	0.1890	0.0960	0.0270
	2	0.0270	0.0960	0.1890	0.2880	0.3750	0.4320	0.4410	0.3840	0.2430
	3	0.0010	0.0080	0.0270	0.0640	0.1250	0.2160	0.3430	0.5120	0.7290
4	0	0.6561	0.4096	0.2401	0.1296	0.0625	0.0256	0.0081	0.0016	0.0001
	1	0.2916	0.4096	0.4116	0.3456	0.2500	0.1536	0.0756	0.0256	0.0036
	2	0.0486	0.1536	0.2646	0.3456	0.3750	0.3456	0.2646	0.1536	0.0486
	3	0.0036	0.0256	0.0756	0.1536	0.2500	0.3456	0.4116	0.4096	0.2916
	4	0.0001	0.0016	0.0081	0.0256	0.0625	0.1296	0.2401	0.4096	0.6561
5	0	0.5905	0.3277	0.1681	0.0778	0.0313	0.0102	0.0024	0.0003	0.0000
	1	0.3280	0.4096	0.3601	0.2592	0.1563	0.0768	0.0284	0.0064	0.0005
	2	0.0729	0.2048	0.3087	0.3456	0.3125	0.2304	0.1323	0.0512	0.0081
	3	0.0081	0.0512	0.1323	0.2304	0.3125	0.3456	0.3087	0.2048	0.0729
	4	0.0005	0.0064	0.0283	0.0768	0.1563	0.2592	0.3601	0.4096	0.3281
	5	0.0000	0.0003	0.0024	0.0102	0.0313	0.0778	0.1681	0.3277	0.5905

n	x	0.1	0.2	0.3	0.4	0.5	0.6	0.7	0.8	0.9
6	0	0.5314	0.2621	0.1176	0.0467	0.0156	0.0041	0.0007	0.0001	0.0000
	1	0.3543	0.3932	0.3025	0.1866	0.0938	0.0369	0.0102	0.0015	0.0001
	2	0.0984	0.2458	0.3241	0.3110	0.2344	0.1382	0.0595	0.0154	0.0012
	3	0.0146	0.0819	0.1852	0.2765	0.3125	0.2765	0.1852	0.0819	0.0146
	4	0.0012	0.0154	0.0595	0.1382	0.2344	0.3110	0.3241	0.2458	0.0984
	5	0.0001	0.0015	0.0102	0.0369	0.0938	0.1866	0.3025	0.3932	0.3543
	6	0.0000	0.0001	0.0007	0.0041	0.0156	0.0467	0.1176	0.2621	0.5314
7	0	0.4783	0.2097	0.0824	0.0280	0.0078	0.0016	0.0002	0.0000	0.0000
	1	0.3720	0.3670	0.2471	0.1306	0.0547	0.0172	0.0036	0.0004	0.0000
	2	0.1240	0.2753	0.3177	0.2613	0.1641	0.0774	0.0250	0.0043	0.0002
	3	0.0230	0.1147	0.2269	0.2903	0.2734	0.1935	0.0972	0.0287	0.0026
	4	0.0026	0.0287	0.0972	0.1935	0.2734	0.2903	0.2269	0.1147	0.0230
	5	0.0002	0.0043	0.0250	0.0774	0.1641	0.2613	0.3177	0.2753	0.1240
	6	0.0000	0.0004	0.0036	0.0172	0.0547	0.1306	0.2471	0.3670	0.3720
	7	0.0000	0.0000	0.0002	0.0016	0.0078	0.0280	0.0824	0.2097	0.4783
8	0	0.4305	0.1678	0.0576	0.0168	0.0039	0.0007	0.0001	0.0000	0.0000
	1	0.3826	0.3355	0.1977	0.0896	0.0313	0.0079	0.0012	0.0001	0.0000
	2	0.1488	0.2936	0.2965	0.2090	0.1094	0.0413	0.0100	0.0011	0.0000
	3	0.0331	0.1468	0.2541	0.2787	0.2188	0.1239	0.0467	0.0092	0.0004
	4	0.0046	0.0459	0.1361	0.2322	0.2734	0.2322	0.1361	0.0459	0.0046
	5	0.0004	0.0092	0.0467	0.1239	0.2188	0.2787	0.2541	0.1468	0.0331
	6	0.0000	0.0011	0.0100	0.0413	0.1094	0.2090	0.2965	0.2936	0.1488
	7	0.0000	0.0001	0.0012	0.0079	0.0313	0.0896	0.1977	0.3355	0.3826
	8	0.0000	0.0000	0.0001	0.0007	0.0039	0.0168	0.0576	0.1678	0.4305

Poisson Probability Tables

These tables provide the probability of exactly x number of occurrences for various values of λ.

Table 2 Poisson Probability Tables

Values of λ

x	0.1	0.2	0.3	0.4	0.5	0.6	0.7	0.8	0.9	1.0
0	0.9048	0.8187	0.7408	0.6703	0.6065	0.5488	0.4966	0.4493	0.4066	0.3679
1	0.0905	0.1637	0.2222	0.2681	0.3033	0.3293	0.3476	0.3595	0.3659	0.3679
2	0.0045	0.0164	0.0333	0.0536	0.0758	0.0988	0.1217	0.1438	0.1647	0.1839
3	0.0002	0.0011	0.0033	0.0072	0.0126	0.0198	0.0284	0.0383	0.0494	0.0613
4	0.0000	0.0001	0.0003	0.0007	0.0016	0.0030	0.0050	0.0077	0.0111	0.0153
5	0.0000	0.0000	0.0000	0.0001	0.0002	0.0004	0.0007	0.0012	0.0020	0.0031
6	0.0000	0.0000	0.0000	0.0000	0.0000	0.0000	0.0001	0.0002	0.0003	0.0005

Values of λ

x	1.1	1.2	1.3	1.4	1.5	1.6	1.7	1.8	1.9	2.0
0	0.3329	0.3012	0.2725	0.2466	0.2231	0.2019	0.1827	0.1653	0.1496	0.1353
1	0.3662	0.3614	0.3543	0.3452	0.3347	0.3230	0.3106	0.2975	0.2842	0.2707
2	0.2014	0.2169	0.2303	0.2417	0.2510	0.2584	0.2640	0.2678	0.2700	0.2707
3	0.0738	0.0867	0.0998	0.1128	0.1255	0.1378	0.1496	0.1607	0.1710	0.1804
4	0.0203	0.0260	0.0324	0.0395	0.0471	0.0551	0.0636	0.0723	0.0812	0.0902
5	0.0045	0.0062	0.0084	0.0111	0.0141	0.0176	0.0216	0.0260	0.0309	0.0361
6	0.0008	0.0012	0.0018	0.0026	0.0035	0.0047	0.0061	0.0078	0.0098	0.0120
7	0.0001	0.0002	0.0003	0.0005	0.0008	0.0011	0.0015	0.0020	0.0027	0.0034
8	0.0000	0.0000	0.0001	0.0001	0.0001	0.0002	0.0003	0.0005	0.0006	0.0009
9	0.0000	0.0000	0.0000	0.0000	0.0000	0.0000	0.0001	0.0001	0.0001	0.0002

Values of λ

x	2.1	2.2	2.3	2.4	2.5	2.6	2.7	2.8	2.9	3.0
0	0.1225	0.1108	0.1003	0.0907	0.0821	0.0743	0.0672	0.0608	0.0550	0.0498
1	0.2572	0.2438	0.2306	0.2177	0.2052	0.1931	0.1815	0.1703	0.1596	0.1494
2	0.2700	0.2681	0.2652	0.2613	0.2565	0.2510	0.2450	0.2384	0.2314	0.2240
3	0.1890	0.1966	0.2033	0.2090	0.2138	0.2176	0.2205	0.2225	0.2237	0.2240
4	0.0992	0.1082	0.1169	0.1254	0.1336	0.1414	0.1488	0.1557	0.1622	0.1680
5	0.0417	0.0476	0.0538	0.0602	0.0668	0.0735	0.0804	0.0872	0.0940	0.1008
6	0.0146	0.0174	0.0206	0.0241	0.0278	0.0319	0.0362	0.0407	0.0455	0.0504
7	0.0044	0.0055	0.0068	0.0083	0.0099	0.0118	0.0139	0.0163	0.0188	0.0216
8	0.0011	0.0015	0.0019	0.0025	0.0031	0.0038	0.0047	0.0057	0.0068	0.0081
9	0.0003	0.0004	0.0005	0.0007	0.0009	0.0011	0.0014	0.0018	0.0022	0.0027
10	0.0001	0.0001	0.0001	0.0002	0.0002	0.0003	0.0004	0.0005	0.0006	0.0008
11	0.0000	0.0000	0.0000	0.0000	0.0000	0.0001	0.0001	0.0001	0.0002	0.0002

Values of λ

x	3.2	3.4	3.6	3.8	4.0	4.2	4.4	4.6	4.8	5.0
0	0.0408	0.0334	0.0273	0.0224	0.0183	0.0150	0.0123	0.0101	0.0082	0.0067
1	0.1304	0.1135	0.0984	0.0850	0.0733	0.0630	0.0540	0.0462	0.0395	0.0337
2	0.2087	0.1929	0.1771	0.1615	0.1465	0.1323	0.1188	0.1063	0.0948	0.0842
3	0.2226	0.2186	0.2125	0.2046	0.1954	0.1852	0.1743	0.1631	0.1517	0.1404
4	0.1781	0.1858	0.1912	0.1944	0.1954	0.1944	0.1917	0.1875	0.1820	0.1755
5	0.1140	0.1264	0.1377	0.1477	0.1563	0.1633	0.1687	0.1725	0.1747	0.1755
6	0.0608	0.0716	0.0826	0.0936	0.1042	0.1143	0.1237	0.1323	0.1398	0.1462
7	0.0278	0.0348	0.0425	0.0508	0.0595	0.0686	0.0778	0.0869	0.0959	0.1044
8	0.0111	0.0148	0.0191	0.0241	0.0298	0.0360	0.0428	0.0500	0.0575	0.0653
9	0.0040	0.0056	0.0076	0.0102	0.0132	0.0168	0.0209	0.0255	0.0307	0.0363
10	0.0013	0.0019	0.0028	0.0039	0.0053	0.0071	0.0092	0.0118	0.0147	0.0181

continues

Values of λ (continued)

x	3.2	3.4	3.6	3.8	4.0	4.2	4.4	4.6	4.8	5.0
11	0.0004	0.0006	0.0009	0.0013	0.0019	0.0027	0.0037	0.0049	0.0064	0.0082
12	0.0001	0.0002	0.0003	0.0004	0.0006	0.0009	0.0013	0.0019	0.0026	0.0034
13	0.0000	0.0000	0.0001	0.0001	0.0002	0.0003	0.0005	0.0007	0.0009	0.0013
14	0.0000	0.0000	0.0000	0.0000	0.0001	0.0001	0.0001	0.0002	0.0003	0.0005
15	0.0000	0.0000	0.0000	0.0000	0.0000	0.0000	0.0000	0.0001	0.0001	0.0002

Cumulative Standard Normal Probability Tables

Table 3 provides the area to the left of the corresponding z-score for the standard normal distribution.

Table 3 Cumulative Standard Normal Probability Tables

Positive z values
Second digit of z

z	0.00	0.01	0.02	0.03	0.04	0.05	0.06	0.07	0.08	0.09
0.0	0.5000	0.5040	0.5080	0.5120	0.5160	0.5199	0.5239	0.5279	0.5319	0.5359
0.1	0.5398	0.5438	0.5478	0.5517	0.5557	0.5596	0.5636	0.5675	0.5714	0.5753
0.2	0.5793	0.5832	0.5871	0.5910	0.5948	0.5987	0.6026	0.6064	0.6103	0.6141
0.3	0.6179	0.6217	0.6255	0.6293	0.6331	0.6368	0.6406	0.6443	0.6480	0.6517
0.4	0.6554	0.6591	0.6628	0.6664	0.6700	0.6736	0.6772	0.6808	0.6844	0.6879
0.5	0.6915	0.6950	0.6985	0.7019	0.7054	0.7088	0.7123	0.7157	0.7190	0.7224
0.6	0.7257	0.7291	0.7324	0.7357	0.7389	0.7422	0.7454	0.7486	0.7517	0.7549
0.7	0.7580	0.7611	0.7642	0.7673	0.7704	0.7734	0.7764	0.7794	0.7823	0.7852
0.8	0.7881	0.7910	0.7939	0.7967	0.7995	0.8023	0.8051	0.8078	0.8106	0.8133
0.9	0.8159	0.8186	0.8212	0.8238	0.8264	0.8289	0.8315	0.8340	0.8365	0.8389
1.0	0.8413	0.8438	0.8461	0.8485	0.8508	0.8531	0.8554	0.8577	0.8599	0.8621
1.1	0.8643	0.8665	0.8686	0.8708	0.8729	0.8749	0.8770	0.8790	0.8810	0.8830
1.2	0.8849	0.8869	0.8888	0.8907	0.8925	0.8944	0.8962	0.8980	0.8997	0.9015

z	0.00	0.01	0.02	0.03	0.04	0.05	0.06	0.07	0.08	0.09
1.3	0.9032	0.9049	0.9066	0.9082	0.9099	0.9115	0.9131	0.9147	0.9162	0.9177
1.4	0.9192	0.9207	0.9222	0.9236	0.9251	0.9265	0.9279	0.9292	0.9306	0.9319
1.5	0.9332	0.9345	0.9357	0.9370	0.9382	0.9394	0.9406	0.9418	0.9429	0.9441
1.6	0.9452	0.9463	0.9474	0.9484	0.9495	0.9505	0.9515	0.9525	0.9535	0.9545
1.7	0.9554	0.9564	0.9573	0.9582	0.9591	0.9599	0.9608	0.9616	0.9625	0.9633
1.8	0.9641	0.9649	0.9656	0.9664	0.9671	0.9678	0.9686	0.9693	0.9699	0.9706
1.9	0.9713	0.9719	0.9726	0.9732	0.9738	0.9744	0.9750	0.9756	0.9761	0.9767
2.0	0.9772	0.9778	0.9783	0.9788	0.9793	0.9798	0.9803	0.9808	0.9812	0.9817
2.1	0.9821	0.9826	0.9830	0.9834	0.9838	0.9842	0.9846	0.9850	0.9854	0.9857
2.2	0.9861	0.9864	0.9868	0.9871	0.9875	0.9878	0.9881	0.9884	0.9887	0.9890
2.3	0.9893	0.9896	0.9898	0.9901	0.9904	0.9906	0.9909	0.9911	0.9913	0.9916
2.4	0.9918	0.9920	0.9922	0.9925	0.9927	0.9929	0.9931	0.9932	0.9934	0.9936
2.5	0.9938	0.9940	0.9941	0.9943	0.9945	0.9946	0.9948	0.9949	0.9951	0.9952
2.6	0.9953	0.9955	0.9956	0.9957	0.9959	0.9960	0.9961	0.9962	0.9963	0.9964
2.7	0.9965	0.9966	0.9967	0.9968	0.9969	0.9970	0.9971	0.9972	0.9973	0.9974
2.8	0.9974	0.9975	0.9976	0.9977	0.9977	0.9978	0.9979	0.9979	0.9980	0.9981
2.9	0.9981	0.9982	0.9982	0.9983	0.9984	0.9984	0.9985	0.9985	0.9986	0.9986
3.0	0.9987	0.9987	0.9987	0.9988	0.9988	0.9989	0.9989	0.9989	0.9990	0.9990
3.1	0.9990	0.9991	0.9991	0.9991	0.9992	0.9992	0.9992	0.9992	0.9993	0.9993
3.2	0.9993	0.9993	0.9994	0.9994	0.9994	0.9994	0.9994	0.9995	0.9995	0.9995
3.3	0.9995	0.9995	0.9995	0.9996	0.9996	0.9996	0.9996	0.9996	0.9996	0.9997
3.4	0.9997	0.9997	0.9997	0.9997	0.9997	0.9997	0.9997	0.9997	0.9997	0.9998
3.5	0.9998	0.9998	0.9998	0.9998	0.9998	0.9998	0.9998	0.9998	0.9998	0.9998
3.6	0.9998	0.9998	0.9999	0.9999	0.9999	0.9999	0.9999	0.9999	0.9999	0.9999

For $z > 3.69$, *use 0.9999*

Negative z values
Second digit of z

z	0.00	0.01	0.02	0.03	0.04	0.05	0.06	0.07	0.08	0.09
-3.6	0.0002	0.0002	0.0001	0.0001	0.0001	0.0001	0.0001	0.0001	0.0001	0.0001
-3.5	0.0002	0.0002	0.0002	0.0002	0.0002	0.0002	0.0002	0.0002	0.0002	0.0002
-3.4	0.0003	0.0003	0.0003	0.0003	0.0003	0.0003	0.0003	0.0003	0.0003	0.0002
-3.3	0.0005	0.0005	0.0005	0.0004	0.0004	0.0004	0.0004	0.0004	0.0004	0.0003
-3.2	0.0007	0.0007	0.0006	0.0006	0.0006	0.0006	0.0006	0.0005	0.0005	0.0005
-3.1	0.0010	0.0009	0.0009	0.0009	0.0008	0.0008	0.0008	0.0008	0.0007	0.0007
-3.0	0.0013	0.0013	0.0013	0.0012	0.0012	0.0011	0.0011	0.0011	0.0010	0.0010
-2.9	0.0019	0.0018	0.0018	0.0017	0.0016	0.0016	0.0015	0.0015	0.0014	0.0014
-2.8	0.0026	0.0025	0.0024	0.0023	0.0023	0.0022	0.0021	0.0021	0.0020	0.0019
-2.7	0.0035	0.0034	0.0033	0.0032	0.0031	0.0030	0.0029	0.0028	0.0027	0.0026
-2.6	0.0047	0.0045	0.0044	0.0043	0.0041	0.0040	0.0039	0.0038	0.0037	0.0036
-2.5	0.0062	0.0060	0.0059	0.0057	0.0055	0.0054	0.0052	0.0051	0.0049	0.0048
-2.4	0.0082	0.0080	0.0078	0.0075	0.0073	0.0071	0.0069	0.0068	0.0066	0.0064
-2.3	0.0107	0.0104	0.0102	0.0099	0.0096	0.0094	0.0091	0.0089	0.0087	0.0084
-2.2	0.0139	0.0136	0.0132	0.0129	0.0125	0.0122	0.0119	0.0116	0.0113	0.0110
-2.1	0.0179	0.0174	0.0170	0.0166	0.0162	0.0158	0.0154	0.0150	0.0146	0.0143
-2.0	0.0228	0.0222	0.0217	0.0212	0.0207	0.0202	0.0197	0.0192	0.0188	0.0183
-1.9	0.0287	0.0281	0.0274	0.0268	0.0262	0.0256	0.0250	0.0244	0.0239	0.0233
-1.8	0.0359	0.0351	0.0344	0.0336	0.0329	0.0322	0.0314	0.0307	0.0301	0.0294
-1.7	0.0446	0.0436	0.0427	0.0418	0.0409	0.0401	0.0392	0.0384	0.0375	0.0367
-1.6	0.0548	0.0537	0.0526	0.0516	0.0505	0.0495	0.0485	0.0475	0.0465	0.0455
-1.5	0.0668	0.0655	0.0643	0.0630	0.0618	0.0606	0.0594	0.0582	0.0571	0.0559
-1.4	0.0808	0.0793	0.0778	0.0764	0.0749	0.0735	0.0721	0.0708	0.0694	0.0681
-1.3	0.0968	0.0951	0.0934	0.0918	0.0901	0.0885	0.0869	0.0853	0.0838	0.0823
-1.2	0.1151	0.1131	0.1112	0.1093	0.1075	0.1056	0.1038	0.1020	0.1003	0.0985
-1.1	0.1357	0.1335	0.1314	0.1292	0.1271	0.1251	0.1230	0.1210	0.1190	0.1170

z	0.00	0.01	0.02	0.03	0.04	0.05	0.06	0.07	0.08	0.09
-1.0	0.1587	0.1562	0.1539	0.1515	0.1492	0.1469	0.1446	0.1423	0.1401	0.1379
-0.9	0.1841	0.1814	0.1788	0.1762	0.1736	0.1711	0.1685	0.1660	0.1635	0.1611
-0.8	0.2119	0.2090	0.2061	0.2033	0.2005	0.1977	0.1949	0.1922	0.1894	0.1867
-0.7	0.2420	0.2389	0.2358	0.2327	0.2296	0.2266	0.2236	0.2206	0.2177	0.2148
-0.6	0.2743	0.2709	0.2676	0.2643	0.2611	0.2578	0.2546	0.2514	0.2483	0.2451
-0.5	0.3085	0.3050	0.3015	0.2981	0.2946	0.2912	0.2877	0.2843	0.2810	0.2776
-0.4	0.3446	0.3409	0.3372	0.3336	0.3300	0.3264	0.3228	0.3192	0.3156	0.3121
-0.3	0.3821	0.3783	0.3745	0.3707	0.3669	0.3632	0.3594	0.3557	0.3520	0.3483
-0.2	0.4207	0.4168	0.4129	0.4090	0.4052	0.4013	0.3974	0.3936	0.3897	0.3859
-0.1	0.4602	0.4562	0.4522	0.4483	0.4443	0.4404	0.4364	0.4325	0.4286	0.4247
-0.0	0.5000	0.4960	0.4920	0.4880	0.4840	0.4801	0.4761	0.4721	0.4681	0.4641

For $z < -3.69$, *use 0.0001*

Student's t-Distribution

Table 4 provides the *t*-statistic for the corresponding value of alpha, or confidence interval, and the number of degrees of freedom.

Table 4 Student's *t*-Distribution

One tail α	0.200	0.150	0.100	0.0500	0.250	0.010	0.005
Two tail α	0.400	0.300	0.200	0.100	0.050	0.020	0.010
Conf lev	0.600	0.700	0.800	0.900	0.950	0.980	0.990
d.f.							
1	1.376	1.963	3.078	6.314	12.706	31.821	63.657
2	1.061	1.386	1.886	2.920	4.303	6.965	9.925
3	0.978	1.250	1.638	2.353	3.182	4.541	5.841
4	0.941	1.190	1.533	2.132	2.776	3.747	4.604
5	0.920	1.156	1.476	2.015	2.571	3.365	4.032
6	0.906	1.134	1.440	1.943	2.447	3.143	3.707

continues

Table 4 Student's t-Distribution (continued)

One tail α	0.200	0.150	0.100	0.0500	0.0250	0.010	0.005
Two tail α	0.400	0.300	0.200	0.100	0.050	0.020	0.010
Conf lev	0.600	0.700	0.800	0.900	0.950	0.980	0.990
d.f.							
8	0.889	1.108	1.397	1.860	2.306	2.896	3.355
9	0.883	1.100	1.383	1.833	2.262	2.821	3.250
10	0.879	1.093	1.372	1.812	2.228	2.764	3.169
11	0.876	1.088	1.363	1.796	2.201	2.718	3.106
12	0.873	1.083	1.356	1.782	2.179	2.681	3.055
13	0.870	1.079	1.350	1.771	2.160	2.650	3.012
14	0.868	1.076	1.345	1.761	2.145	2.624	2.977
15	0.866	1.074	1.341	1.753	2.131	2.602	2.947
16	0.865	1.071	1.337	1.746	2.120	2.583	2.921
17	0.863	1.069	1.333	1.740	2.110	2.567	2.898
18	0.862	1.067	1.330	1.734	2.101	2.552	2.878
19	0.861	1.066	1.328	1.729	2.093	2.539	2.861
20	0.860	1.064	1.325	1.725	2.086	2.528	2.845
21	0.859	1.063	1.323	1.721	2.080	2.518	2.831
22	0.858	1.061	1.321	1.717	2.074	2.508	2.819
23	0.858	1.060	1.319	1.714	2.069	2.500	2.807
24	0.857	1.059	1.318	1.711	2.064	2.492	2.797
25	0.856	1.058	1.316	1.708	2.060	2.485	2.787
26	0.856	1.058	1.315	1.706	2.056	2.479	2.779
27	0.855	1.057	1.314	1.703	2.052	2.473	2.771
28	0.855	1.056	1.313	1.701	2.048	2.467	2.763
29	0.854	1.055	1.311	1.699	2.045	2.462	2.756
30	0.854	1.055	1.310	1.697	2.042	2.457	2.750

APPENDIX

C

Glossary

Addition Rule of Probabilities Determines the probability of the union of two or more events.

Alternative Hypothesis Denoted by H_1, represents the opposite of the null hypothesis and holds true if the null hypothesis is found to be false.

Analysis of Variance (ANOVA) A procedure to test the difference between more than two population means.

Bar Chart A data display where the value of the observation is proportional to the height of the bar on the graph.

Bayes' Theorem A theorem used to calculate P(A|B) from information about P(B|A). The term P(B|A) refers to the probability of Event B, given that Event A has occurred.

Biased Sample A sample that does not represent the intended population and can lead to distorted findings.

Binomial Experiment An experiment that has only two possible outcomes for each trial. The probability of success and failure is constant. Each trial of the experiment is independent of any other trial and there is a fixed number of trials.

Binomial Probability Distribution A method used to calculate the probability of a specific number of successes for a certain number of trials.

Central Limit Theorem A theorem that states as the sample size, *n*, gets larger, the sampling distribution of the sample means tends to follow a normal probability distribution with mean equal to μ and standard error equal to $\frac{\sigma}{\sqrt{n}}$.

Class The interval in a frequency distribution.

Class Midpoint The point in the middle of each class and is calculated as the sum of the lower limit and the upper limit divided by 2.

Classical Probability An approach to calculate the probability of an event in situations when we know the number of possible outcomes of the event of interest.

Cluster Sample A simple random sample of groups, or clusters, of the population. Each member of the chosen clusters would be part of the final sample.

Coefficient of Determination, R^2 A measure of goodness-of-fit for the regression model and represents the percentage of the variation in y that is explained by the variation in x.

Combinations The number of different ways in which objects can be arranged without regard to order.

Completely Randomized One-Way ANOVA An analysis of variance procedure that involves the independent random selection of observations for each level of one factor.

Conditional Probability The probability that Event A will occur, knowing that Event B has already occurred, written as $P(A|B)$.

Confidence Interval A range of values used to estimate a population parameter and is associated with a specific confidence level.

Confidence Level The probability that the interval estimate will include the population parameter.

Contingency Table A table which shows the actual or relative frequency of two variables at the same time.

Continuous Random Variable A variable that can assume any numerical value within an interval, and results from measuring the outcome of an experiment.

Correlation Coefficient Indicates the strength and direction of the linear relationship between the independent and dependent variables.

Cumulative Frequency Distribution Indicates the number of observations that are less than or equal to the current class.

Data The value assigned to an observation or a measurement and the building block to statistical analysis.

Degrees of Freedom The number of values that are free to be varied given information, such as the sample mean, is known.

Dependent Samples The observation from one sample is related to an observation from another sample.

Dependent Variable The variable denoted by y in the regression equation that is influenced by the independent variable. It is also known as the response variable or the explained variable.

Descriptive Statistics Used to summarize or display data in a more meaningful way so that an overview can quickly be obtained.

Direct Observation Gathering data while the subjects of interest are in their natural environment.

Discrete Probability Distribution A listing of all the possible outcomes of an experiment for a discrete random variable along with the relative frequency or probability.

Discrete Random Variable A variable that is limited to assuming only specific integer values and results from counting the outcome of an experiment.

Empirical Probability Type of probability that observes the number of occurrences of an event through an experiment and calculates the probability from a relative frequency distribution.

Empirical Rule If a distribution follows a bell-shaped, symmetrical curve centered around the mean, then we would expect approximately 68.3, 95.5, and 99.7 percent of the values to fall within one, two, and three standard deviations around the mean, respectively.

Event One or more outcomes that are of interest for the experiment and which is/are a subset of the sample space.

Expected Frequencies The number of observations that would be expected for each category of a frequency distribution, assuming the null hypothesis is true with chi-square analysis.

Experiment The process of measuring or observing an activity for the purpose of collecting data.

Factor Describes the cause of the variation in the data for analysis of variance.

Focus Group An observational technique where the subjects are aware that data is being collected, often in the format of guided interviews. Businesses use this type of group to gather information in a group setting that is controlled by a moderator.

Frequency Distribution A table that shows the classes and number of data observations that fall into each class.

Fundamental Counting Principle A concept that states if one event can occur in m ways and a second event can occur in n ways, then the total number of ways both events can occur together is $m \cdot n$ ways. This can be extended to more than two events.

Goodness-of-Fit Test Uses a sample to test whether a frequency distribution fits the predicted distribution.

Histogram A bar graph representing the classes on the horizontal axis and the number of observations in each class on the vertical axis.

Hypothesis An assumption about a population parameter.

Hypothesis Testing A procedure that uses sample information to test an assumption about a population parameter.

Independent Events The occurrence of Event B has no effect on the probability of Event A.

Independent Samples The observation from one sample is not related to any observations from another sample.

Independent Variable The variable denoted by x in the regression equation that is suspected to influence the dependent variable. It is also known as the control variable or the explanatory variable.

Inferential Statistics Used to make claims or conclusions about a population based on a sample of data from that population.

Interquartile Range Measures the spread of the center half of the data set and is used to identify outliers.

Intersection Two or more events occurring at the same time.

Interval Estimate Provides a range of values that best describe the population.

Interval Level of Measurement Level of data that allows the use of addition and subtraction when comparing values, but the zero point is arbitrary.

Joint Probability The probability of the intersection of two events.

Law of Large Numbers This law states that when an experiment is conducted a large number of times, the empirical probabilities of the process will converge to the classical probabilities.

Level The number of categories within the factor of interest in the analysis of variance procedure.

Level of Significance (α) Probability of making a Type I error.

Line Chart A display where ordered pair data points are connected together with a line.

Margin of Error Concept that determines the width of a confidence interval and is calculated using $z_{\alpha/2}\sigma_{\bar{X}}$.

Mean A measure of the central point and is calculated by adding all the values in the data set and then dividing this result by the number of observations.

Mean Square Between (MSB) A measure of variation between the sample means.

Mean Square Within (MSW) A measure of variation within each sample.

Measure of Central Tendency Describes the center point of a data set with a single value.

Measure of Relative Position Describes the percentage of the data below a certain point.

Median The value in the data set for which half the observations are higher and half the observations are lower.

Mode The observation in the data set that occurs most frequently.

Multiplication Rule of Probabilities This rule determines the probability of the intersection of two or more events.

Mutually Exclusive Events When two events cannot occur at the same time during an experiment.

Nominal Level of Measurement Lowest level of data where numbers are used to identify a group or category.

Normal Probability Distribution A continuous probability distribution that is symmetrical, bell-shaped, and asymptotic.

Null Hypothesis Denoted by H_0, this represents the case that there is no difference between the population and the sample and involves stating the belief that the population parameter is $\leq$, $=$, or $\geq$ a specific value.

Observed Frequencies The number of actual observations noted for each category of a frequency distribution with chi-square analysis.

Observed Level of Significance The smallest level of significance at which the null hypothesis will be rejected, assuming the null hypothesis is true. It is also known as the *p*-value.

One-Tail Hypothesis Test This test is used when the rejection area is in only one tail of the distribution and when the alternative hypothesis is stated as $<$ or $>$.

One-Way ANOVA An analysis of variance procedure where only one factor is being considered.

Ordinal Level of Measurement This measurement has all the properties of nominal data with the added feature that we can rank the values from highest to lowest.

Ordinary Least Squares Method A mathematical procedure to identify the linear equation that best fits the data for two variables x and y by finding values for a, the y-intercept; and b, the slope. The goal of the ordinary least squares method is to minimize the sum of the squared difference between the values of y and $\hat{y}$.

Outcome A particular result of an experiment.

Outliers Extreme values in a data set that should be examined before being used in any analysis.

***p*-Value** The smallest level of significance at which the null hypothesis will be rejected, assuming the null hypothesis is true.

Parameter Data that describes a characteristic about a population.

Percentiles Measures of the relative position of the data values from dividing the data set into 100 equal segments.

Permutations The number of different ways in which objects can be arranged in order.

Pie Chart A chart used to describe data from relative frequency distributions with a circle divided into portions whose area is equal to the relative frequency distribution.

Point Estimate A single value that best describes the population of interest, the sample mean being the most common.

Poisson Probability Distribution A measurement that is used to calculate the probability that a certain number of events will occur over a specific period of time or space.

Pooled Estimate of the Variance A weighted average of two sample variances.

Population A number which represents all possible outcomes or measurements of interest.

Primary Data Data that is collected by the person who eventually uses the data.

Probability A measure of likelihood that a particular event will occur.

Probability Distribution A listing of all the possible outcomes of an experiment along with the relative frequency or probability of each outcome.

Qualitative Variable A variable with a non-numerical value such as gender or race.

Quantitative Variable A variable with a numerical value such as age or income.

Quartiles Measures the relative position of the data values by dividing the data set into four equal segments.

Random Variable A variable that takes on a numerical value as a result of an experiment.

Randomized Block ANOVA Analysis of variance procedure that controls for variations from sources other than the factors of interest.

Range Obtained by subtracting the smallest measurement from the largest measurement of a sample.

Ratio Level of Measurement Level of data that allows the use of all four mathematical operations to compare values and has a true zero point.

Relative Frequency Distribution Displays the percentage of observations of each class relative to the total number of observations.

Residual The difference between the actual value of y and the estimated value of $\hat{y}$.

Sample A subset of a population.

Sample Space All the possible outcomes of an experiment.

Sampling Distribution for the Difference in Means Describes the probability of observing various intervals for the difference between two sample means.

Sampling Distribution of the Mean The pattern of the sample means that will occur as samples are drawn from the population at large.

Sampling Error An error that occurs when the sample measurement is different from the population measurement.

Scheffé Test A test used to determine which of the sample means are different after rejecting the null hypothesis using analysis of variance.

Secondary Data Data that somebody else has collected and made available for others to use.

Simple Linear Regression A procedure that describes a straight line that best fits the data for two variables x and y.

Simple Random Sample A sample where every element in the population has a chance at being selected.

Standard Deviation A measure of variation calculated by taking the square root of the variance.

Standard Error of the Difference Between Two Means The error describes the variation in the difference between two sample means.

Standard Error of the Mean The standard deviation of sample means.

Standard Error of the Proportion The standard deviation of the sample proportions.

Standard Error of the Regression (s_e) Measures the amount of dispersion of the observed data around the estimated regression line.

Standard Normal Distribution A normal distribution with a mean equal to zero and standard deviation equal to one.

Statistic Data that describes a characteristic about a sample.

Statistics The science that deals with the collection, tabulation, presentation, and systematic classification of quantitative data, especially as a basis for inference and induction.

Stem and Leaf Design A chart that displays the frequency distribution by splitting the data values into leaves (the last digit in the value) and stems (the remaining digits in the value).

Stratified Sample A sample that is obtained by dividing the population into mutually exclusive groups, or strata, and randomly sampling from each of these groups.

Subjective Probability Probability that is estimated based on experience and intuition.

Sum of Squares Between (*SSB*) The variation among the samples in analysis of variance.

Sum of Squares Block (*SSBL*) The variation among the blocks in analysis of variance.

Sum of Squares Within (*SSW*) The variation within the samples in analysis of variance.

Surveys Data collection that involves asking the subjects a series of questions.

Systematic Sample A sample where every kth member of the population is chosen for the sample, with value of k being approximately $\frac{N}{n}$, where N equals the size of the population and n equals the size of the sample.

Test Statistic A quantity from a sample used to decide whether or not to reject the null hypothesis.

Total Sum of Squares The total variation in analysis of variance that is obtained by adding the sum of squares between (*SSB*) and the sum of squares within (*SSW*).

Two-Tail Hypothesis Test This test is used when the alternative hypothesis is expressed as $\neq$ and when there are two areas of rejection, one in each tail.

Type I Error Occurs when the null hypothesis is rejected when, in reality, it is true.

Type II Error Occurs when the null hypothesis is not rejected when, in reality, it is not true.

Union At least one of a number of possible events occur.

Variance A measure of dispersion that describes the relative distance between the data points in the set and the mean of the data set.

Weighted Mean Measure which allows the assignment of more weight to certain values and less weight to others when calculating an average.

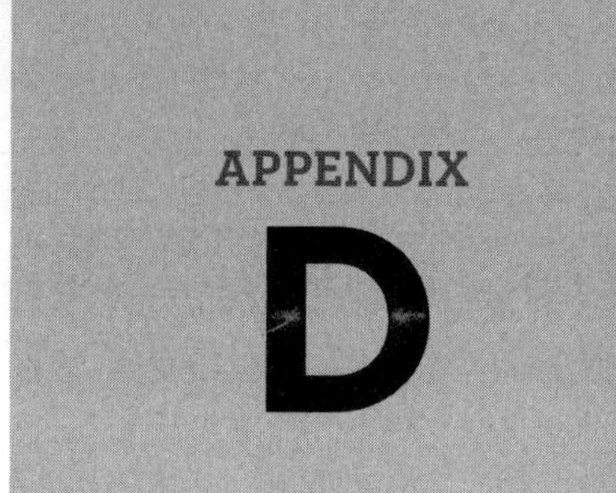

Statistics Reference Glossary

Descriptive Statistics

Population Mean	$\mu = \frac{\sum_{i=1}^{N} x_i}{N}$	Population Variance	$\sigma^2 = \frac{\sum_{i=1}^{N} (x_i - \mu)^2}{N}$
Sample Mean	$\bar{x} = \frac{\sum_{i=1}^{n} x_i}{n}$	Sample Variance	$s^2 = \frac{\sum_{i=1}^{n} (x_i - \bar{x})^2}{n-1}$

Probability Distributions

Mean	$\mu = \sum_{i=1}^{n} x_i P(x_i)$	Variance	$\sigma^2 = \sum_{i=1}^{n} (x_i - \mu)^2 P(x_i)$
Binomial Probability Distribution	$P(x) = \frac{n!}{x!(n-x)!} p^x q^{(n-x)}$		
Poisson Probability Distribution	$P(x) = \frac{\lambda^x e^{-\lambda}}{x!}$		

Confidence Intervals

Type	σ	Population	Sample	Confidence Interval
Mean	Known	Any	$n \geq 30$	$\bar{x} \pm z_{\alpha/2} \frac{\sigma}{\sqrt{n}}$
Mean	Known	Must Be Normal	$n < 30$	$\bar{x} \pm z_{\alpha/2} \frac{\sigma}{\sqrt{n}}$
Mean	Unknown	Any	$n \geq 30$	$\bar{x} \pm z_{\alpha/2} \frac{s}{\sqrt{n}}$
Mean	Unknown	Must Be Normal	$n < 30$	$\bar{x} \pm t_{\alpha/2} \frac{s}{\sqrt{n}}$
Proportion		Any	$np \geq 5$	$\bar{p} \pm z_{\alpha/2} \sqrt{\frac{\bar{p}(1-\bar{p})}{n}}$

Sample Size for Confidence Intervals

Type	Sample Size
Mean	$n=\left(\frac{z\,\sigma}{ME}\right)^2$
Proportion	$n=pq\left(\frac{z_{\alpha/2}}{ME}\right)^2$

Critical z-Scores

Alpha	Tail	Critical Z-Score	Alpha	Tail	Critical Z-Score
0.01	Two	±2.58	0.01	One	±2.33
0.05	Two	±1.96	0.05	One	±1.65
0.10	Two	±1.65	0.10	One	±1.28

One-Sample Hypothesis Tests

Type	σ	Population	Sample	Test Statistic
Mean	Known	Any	$n \geq 30$	$z=\frac{\bar{x}-\mu}{\sigma/\sqrt{n}}$
Mean	Known	Must Be Normal	$n < 30$	$z=\frac{\bar{x}-\mu}{\sigma/\sqrt{n}}$
Mean	Unknown	Any	$n \geq 30$	$z=\frac{\bar{x}-\mu}{s/\sqrt{n}}$
Mean	Unknown	Must Be Normal	$n < 30$	$t=\frac{\bar{x}-\mu}{s/\sqrt{n}}$ $d.f.=n-1$
Proportion		Any	$np \geq 5$	$z=\frac{\bar{p}-p}{\sqrt{\frac{\bar{p}(1-\bar{p})}{n}}}$

Two-Sample Hypothesis Tests

Type	σ_1 , σ_2	Population	Sample	Test Statistic
Mean	Known	Any	$n_1, n_2 \geq 30$ Independent Samples	$z = \frac{(\bar{x}_1 - \bar{x}_2) - (\mu_1 - \mu_2)}{\sqrt{\frac{\sigma_1^2}{n_1} + \frac{\sigma_2^2}{n_2}}}$
Mean	Known	Must Be Normal	$n_1, n_2 < 30$ Independent Samples	$z = \frac{(\bar{x}_1 - \bar{x}_2) - (\mu_1 - \mu_2)}{\sqrt{\frac{\sigma_1^2}{n_1} + \frac{\sigma_2^2}{n_2}}}$
Mean	Unknown	Any	$n_1, n_2 \geq 30$ Independent Samples	$z = \frac{(\bar{x}_1 - \bar{x}_2) - (\mu_1 - \mu_2)}{\sqrt{\frac{s_1^2}{n_1} + \frac{s_2^2}{n_2}}}$
Mean	Unknown and Equal	Must Be Normal	$n_1, n_2 < 30$ Independent Samples	$t = \frac{(\bar{x}_1 - \bar{x}_2) - (\mu_1 - \mu_2)}{\sqrt{\frac{(n_1 - 1)s_1^2 + (n_2 - 1)s_2^2}{n_1 + n_2 - 2}} \sqrt{\frac{1}{n_1} + \frac{1}{n_2}}}$ $d.f. = n_1 + n_2 - 2$
Mean	Unknown and Unequal	Must Be Normal	$n_1, n_2 < 30$ Independent Samples	$t = \frac{(\bar{x}_1 - \bar{x}_2) - (\mu_1 - \mu_2)}{\sqrt{\frac{s_1^2}{n_1} + \frac{s_2^2}{n_2}}}$ $d.f. = \frac{\left(\frac{s_1^2}{n_1} + \frac{s_2^2}{n_2}\right)^2}{\frac{\left(\frac{s_1^2}{n_1}\right)^2}{n_1 - 1} + \frac{\left(\frac{s_2^2}{n_2}\right)^2}{n_2 - 1}}$
Proportion		Any	$np \geq 5$ $nq \geq 5$ Independent Samples	$z = \frac{(\bar{p}_1 - \bar{p}_2) - (p_1 - p_2)_{H_0}}{\sqrt{(\hat{p})(1 - \hat{p})\left(\frac{1}{n_1} + \frac{1}{n_2}\right)}}$

APPENDIX

E

Excel Reference Glossary

Probability Distributions

FACT(*n*)	Factorial
PERMUT(*n*, *r*)	Permutations
COMBIN(*n*, *r*)	Combinations
BINOM.DIST(*x*, *n*, *p*, cumulative)	Binomial probability distribution
POISSON.DIST(*y*, λ, cumulative)	Poisson probability distribution
NORM.DIST(*x*, mean, standard_dev, cumulative)	Normal probability distribution

Confidence Intervals

CONFIDENCE.NORM(alpha, standard_dev, size)	Confidence interval using normal distribution
CONFIDENCE.T(alpha, standard_dev, size)	Confidence interval using the *t*-distribution

One-Sample Hypothesis Tests

NORM.S.DIST(*z*, cumulative)	The *p*-value for a calculated *z*-test statistic
T.DIST(*x*, Deg_freedom)	The *p*-value for a calculated *t*-test statistic, left-tail test
T.DIST.RT(*x*, Deg_freedom)	The *p*-value for a calculated *t*-test statistic, right-tail test
T.DIST.2T	The *p*-value for a calculated *t*-test statistic, two-tail test
T.INV(Probability, Deg_freedom)	The critical *t*-value for a one-tail test
T.INV.2T(Probability, Deg_freedom)	The critical *t*-value for a two-tail test

Two-Sample Hypothesis Tests (Data Analysis ToolPak)

z-Test: Two Samples for Means	*z* test when σ_1 and σ_2 are known
t-Test: Two-Sample Assuming Equal Variances	*t*-test, σ_1 and σ_2 are unknown, independent samples, and equal population variances
t-Test: Two-Sample Assuming Unequal Variances	*t*-test, σ_1 and σ_2 are unknown, independent samples, unequal population variances
t-Test: Paired Two Sample for Means	Hypothesis test with dependent samples

Index

Symbols

A

B

C

D

E

F

G

H

I

J-K-L

M

N

O

P

Q

R

S

T

U

V

W-X-Y-Z